ALL WHAT MATTERS

The Facts about Global Climate Genesis

silence

write

speak

Uli Weber

About this book

The progress in the natural sciences had once inspired the Western enlightenment. Their domestication through externally funded courtesy studies and an increasing ideologization up to hysterical revival movements may therefore usher in another dark age, namely the obsessive decarbonization craze up to the year 2100.

In the 1970s, the author was taught the basic principles of paleoclimatology. He therefore considered the early climate alarm of the 1980s to be a not entirely unselfish media focus on extremes in the spectrum of honest scientific knowledge. In the 1990s, he missed a bridge between current climate research and secured paleoclimatology, and in the 2000s, considerable contradictions emerged between the two. In the 2010s, the author then had the time to seriously examine the scientific arguments of the current climate hysteria. He first challenged the consistency of these foundations and then refuted them on the basis of established scientific laws.

Despite various publications, the refutations of the climate alarmist core theses summarized herein did not find any place in the (German) public climate discussion. Talking and writing is all what is possible, and the silence has not been neglected. In the course of his sociopolitical efforts the author was able to realize a certain loss in interpersonal contacts; he also considers the ad hominem appraisals from STEM-distant fellow realists, triggered by his public contributions, to be dispensable.

Unfortunately, the author had not received any sponsorship for the work presented in this book - but he would consider 400ppm of the funding for scientific climate hysteria to be a reasonable share ...

Real Science is Committed to Truth Alone

To put it simply, a current quote sums it up: "*This means that we do not call lies truths and truths not lies!*"

(German Chancellor Dr. Dr. h.c. Angela Merkel on 05/30/2019 @ Harvard)

silence

write

speak

ALL WHAT MATTERS - The Facts about Global Climate Genesis

© 2020 Uli Weber

Production and publishing:

BoD - Books on Demand GmbH, Norderstedt

ISBN: 978-3-75288-703-7

Cover Design: U. Weber 2019 (inspired by Tucholskys „Stairs")

Acknowledgments: The author would like to thank the German Geophysical Society most sincerely and in deep respect for the straightforward scientific stance with which the DGG editorial team had published his papers and put it up for discussion.

Hamburg on June 1, 2019 - Uli Weber

This book is an English translation of "MEHR GEHT NICHT - Ein klimawissenschaftliches Vermächtnis", published July 10, 2019 (ISBN 9783744818513).

The author must apologize to the valued reader for his basic use of terms and expressions in English and for possible mistakes in the translation of his German text.

Hamburg on February 5, 2020 - Uli Weber

Bibliographic information from the German National Library:

The German National Library lists this publication in the German National Bibliography; detailed bibliographical data can be found on the Internet at (http://dnb.d-nb.de).

Content

The Beginnings of the Climate Religion

At the end of the 1980s / early 1990s at the latest, the current hysteria surrounding a man-made climate catastrophe had started in very different places and then quickly developed into a global political agenda:

IPCC 1988, quote from German Wikipedia (status 5-2019):

„The Intergovernmental Panel on Climate Change (IPCC, Intergovernmental Committee on Climate Change), often referred to in German as the "world climate council", was designated in November 1988 by the United Nations Environment Program (UNEP) and the World Meteorological Organization (WMO) as an intergovernmental Institution to summarize the state of scientific research on climate change for political decision-makers with the aim of providing the basis for science-based decisions without giving any recommendations for action."

German Bundestag 1989, quote from the Drucksache 11/4133 dated 08th of March, 1989:

„Stratospheric ozone depletion and the greenhouse effect are becoming an ever greater challenge for mankind. The threat to the earth's atmosphere endangers life on earth if current developments are not stopped early and comprehensively. The cause of the danger are trace gases released by human activities."

(First recommendation for a resolution and report by the Committee on the Environment, Nature Conservation and Nuclear Safety on the first interim report of the Enquete Commission "Precautionary measures to protect the earth's atmosphere" in accordance with the decision of the German Bundestag of October 16 and November 27, 1987 - printed matter 11 / 533, 11/787, 11/971, 11/1351, 11/3246)

Club of Rome 1991: The book "The First Global Revolution" (1991) by Alexander King and Bertrand Schneider for the Club of Rome states on page 70, quotation:

> *„The need for enemies seems to be a common historical factor. Some states have striven to overcome domestic failure and internal contradictions by blaming external enemies. The ploy of finding a scapegoat is as old as mankind itself - when things become too difficult at home, divert attention to adventure abroad. Bring the divided nation together to face an outside enemy, either a real one, or else one invented for the purpose."*

And at page 75 there is frankly stated, quotation:

> *„In searching for a common enemy against whom we can unite, we came up with the idea that pollution, the thread of global warming, water shortages famine and the like, would fit the bill. In their totality and their interactions these phenomena do constitute a common thread which must be confronted by everyone together. But in designating these dangers as the enemy, we fall into the trap, which we have already warned readers about, namely mistaking symptoms for causes.*
> *All these dangers are caused by human intervention in natural processes, and it is only through changed attitudes and behaviour that they can be overcome. The real enemy than is humanity itself."*

The Framework Convention on Climate Change (UNFCCC) from 1992, quotation from German Wikipedia (Stand 5-2019):

> *„In June 1992, the United Nations Conference on Environment and Development (UNCED) was held in Rio de Janeiro. At the time, the world's largest international conference, delegates from almost all governments and representatives of numerous non-governmental organizations traveled to Brazil. Several multilateral environmental agreements have been agreed in Rio, including the United Nations Framework Con-*

vention on Climate Change (UNFCCC). In addition, Agenda 21 should promote the increased efforts for more sustainability, particularly at regional and local level, which from then on included climate protection."

The Kyoto-Protokoll from 1997, quotation from German Wikipedia (Stand 5-2019):

„The Kyoto Protocol to the United Nations Framework Convention on Climate Change (short: Kyoto Protocol, named after the location of the Kyoto conference in Japan) is an additional protocol adopted on December 11, 1997 for the design of the United Nations Framework Convention on Climate Change (UNFCCC) the goal of climate protection."

The 2015 UN Climate Change Conference in Paris then produced a non-binding global climate deal and an annual $ 100 billion redistribution program for emerging and developing countries, leaving it unclear what the egg and what the hen may have been.

The offer in Paris was apparently, for a signature under the non-binding climate contract there is an annual patronage from the Green Climate Fund (GCF). Can a responsible politician from a poor country really say "no" to this? - In the book "The Godfather" it was said at such a point, *"we make him an offer that he cannot refuse ..."*.

As a result, we can conclude that the industrialized nations in their climate mania obviously pay the emerging countries and the Third World to agree to decarbonization of the world without making their own commitment. Conversely, such a climate imperialism will also ensure that these forced-missioned countries can be blackmailed permanently by increasing financial dependency and prevent any contrary opinion from there.

If we rummage a bit further in the German Wikipedia as of June 1, 2019, there will be more interesting connections. The following Wikipedia quotes have been partially shortened and summarized:

Earth Day, which starts every year on April 22nd in over 175 countries, is designed to raise awareness of the natural environment and encourage people to rethink the nature of consumer behavior.

And Vladimir Ilyich Ulyanov, called Lenin, was born in Simbirsk on April 22, 1870, according to the Gregorian calendar.

After the failed UN climate conference in Copenhagen in 2009, Bolivian President Evo Morales hosted International Mother Earth Day 2010 and an alternative world conference for people on climate change and Mother Earth rights.

In his opening speech, President Morales identified the capitalist system as the main cause of the imbalance on Earth, since the planet and its inhabitants would suffer from the boundless growth constraint. A ten-page "Peoples Agreement" was drawn up in 17 working groups that met in crowded lecture halls. One of the main causes of climate change is the agricultural sector, which produces food for the market but not for the nutrition of all people. The industrialized countries are encouraged to halve their CO2 emissions by 2020 and to pay six percent of their annual budget into a global climate fund. Companies and governments should be able to be sued in front of a world climate court to be founded. A global referendum on environmental protection is to be organized jointly by governments, environmental organizations and trade unions. ... Since then, "Earth Day" has been called "International Mother Earth Day".

Movimiento al Socialismo (MAS; más means "more" in Spanish) is the name of a leftist party in Bolivia led by Evo Morales. With Evo Morales she has been the President of Bolivia since the end of 2005.

So it should come as no surprise if the rise of the climate religion coincides with the fall of the Iron Curtain. The events described here from several decades can neither be recognized in their development nor in their results as a possible concerted action by a "responsible citizen". In retrospect, however, these connections are quite suitable to support a "climate-communist conspiracy theory" for the global redistribution of the wealth of western industrial nations; maybe the worshipers of Lenin & Co. simply defected to Mother Gaia and therefore celebrate on April 22 each year their globalization fantasies...

Because all so-called scientific facts on which these global climate agreements are based are wrong:

The author initially had only two basic problems of understanding the paradigm of man-made climate change and also missed a specific statement.

Firstly, he had accepted the so-called "natural atmospheric greenhouse effect" (THE) as secured scientific knowledge, but never really understood it, because it conflicts with the 2nd law of thermodynamics.

Secondly, he was unable to understand scientifically the theory of so-called "greenhouse gases" (GHG) because these "greenhouse gases" are not able to generate additional energy by themselves. Rather, energy is simply passed on here in the form of a relay race. Or does the valued reader perhaps have a pure CO_2 heater in his house that does not consume any energy?

No?

But if CO_2 cannot heat at all, then the problem of a "man-made" climate catastrophe is obsolete.

And finally, there was still no scientific information about the absolute share of natural paleo climate forcing as part of the current climate change. Another point was added in the course of his research: There is no 97% consensus in science about a "man-made" climate catastrophe.

There are two basic approaches to a scientific review. On the one hand, one can assume that the paradigm in question is correct and check its consequences based on reality; on the other hand, one can also compare the theoretical basis of this paradigm with the established scientific laws. The author had therefore first tried to understand the statements of climate science from their own paradigm and came to the conclusion that these basic assumptions cannot be used for the current climate alarm even if they were correct:

An alleged climate forcing from CO_2 would be well below 1 degree Celsius per doubling of CO_2.

A global CO_2 budget would be continuously renewable.

Because the so-called GHG can only absorb existing radiation energy and release it again, whereby they can only release what they had previously absorbed. So they are just "ping pong players" in a physical zero-sum game. Furthermore, the topicality-principle of geology requires that the natural (paleo-) climatic laws do not change arbitrarily in the course of the earth's history:

> CO_2 is not the natural climate driver on our earth.

> A sole solar climate forcing via the albedo of the earth is computationally possible and is outlined here below.

And finally there is no "natural atmospheric greenhouse effect"; this construct only conceals a grossly incorrect application of the physical Stefan-Boltzmann law. This statement may disturb the reader, so here is an example from real life: Let's say you go regularly to a tanning salon. One day they will make you a great offer for a super-modern turbo tanner. Standing on a turntable, you are slowly rotated and receive twice the amount of radiation from a stationary radiator as before. The average amount of radiation over the entire area of your body would remain the same, but the tanning time would be halved because the radiation was doubled.

Would you really accept such an offer?

Perhaps you would attribute the functionality of such a turbo tanner to the principle of a chicken grill, the heating output of which is simply calculated down to the whole chicken, including its back. And so you could come to the realization that the temperature of twice as much direct radiation would not be good for your health.

According to this "chicken grill tanner", the other way round, the value for the actual "natural" temperature of our earth has been calculated far too low. One simply distributes the actual sun radiation on the day side mathematically over the entire surface of the earth, i.e. also on its night side. And then people say that the sun is not strong enough to generate the measured temperatures on our earth with its direct radiation. In conclusion, this "chicken trick" has been used inversely to demonstrate a supposedly "natural atmospheric greenhouse effect" which has never been proved in reality.

Direct sunlight on our earth can generate a maximum temperature of around 120°C at the equatorial zenith. Less the average reflected radiation component and by averaging over the entire earth, however, according to conventional calculations, the result is only -18°C. In relation to the chicken grill, where you also average the output of the radiant heater on the entire chicken including its back, this would make the difference between a grilled (120°C) and a frozen (-18°C) chicken...

The fundamental mistake in thinking when calculating a "natural" global temperature lies in the distribution of solar radiation over the entire surface of the earth. The introduction of an also incorrect correction factor called "natural atmospheric greenhouse effect" does not result in a correct scientific solution, but only enforces a foremost agreement with the observed reality. To date, there is not a single scientific experiment that can prove this incorrect correction factor.

And the next time you'll pass a grill room, just take a look inside and convince yourself...

The Wrong 97% Consensus in Science

We are all individuals, i.e. individual cases. Have you ever been the first to complain about anything? The author made the experience that in such cases he gradually became aware of a number of other "first complaints". Such "first complaints" are the ultimate means of power of the respective complaints office to prove that they "knew nothing". We assume that this complaints office has a complete overview of all complaints received, and the reverse, then, sets us up against the wall as an individual case; because you don't have to worry about isolated "cross drivers" there...

Or have you heard the other way around that many others have already complained about your complaint?

Majorities and minorities are always associated with evaluations in our everyday experience, which we use to intuitively orient ourselves. Conversely, through the mediation of such majority or minority opinions, we can be "directed" to a desired thinking or behavior; just have a look in the DigitalWiki under the term "nudging".

Imagine you would get from the media that 97% of all scientists confirm the "man-made climate change". Of course, you assume that such a message has been editorially checked beforehand. And you will definitely not research the scientific publication that may have been cited there.

But there is no 97% consensus in science!

Blog article: *„Das siebenundneunzig Prozent-Problem: Welcher Konsens?" (U. Weber)*

Published in German at KalteSonne Blog on February 19, 2015:
http://www.kaltesonne.de/das-siebenundneunzig-prozent-problem-welcher-konsens/

An English translation was published at NoTricksZone on February 20, 2015:
http://notrickszone.com/2015/02/20/german-analysis-97-percent-consensus-does-not-exist-demands-to-end-debate-are-way-off-sides/#sthash.J67G1j00.dpbs

The ninety-seven percent problem: what consensus?

Again and again one hears and reads that 97 percent of all scientific work (sometimes also of all scientists) would confirm man-made global warming. In this statement, the Consensus Project even refers to a published study that wants to have proven exactly that. The study "Quantifying the consensus on anthropogenic global warming in the scientific literature" by Cook et al. from Environ. Res. Lett. 8 (2013) 024024 (7pp) demonstrates the 97% consensus for "Anthropogenic Global Warming" (AGW) as follows:

- 12,465 scientific papers were examined for statements about AGW

- 4,014 papers contained their own positions on AGW

- Of these 4,014 papers with statements about AGW, 97% confirm the AGW theory

The alleged AGW consensus of 97 percent is thus calculated as a circular reference within a subset of 4,014 of the 12,465 scientific works originally examined and not on the basis of the totality of all works. This calculation approach is of course completely absurd and does not gain any meaningfulness. If a statement on AGW can be presented in such a form, then the so-called "consensus" would, if correctly calculated, come to a quota of just under 32% of the scientific work examined. This almost a third of all 12,465 works examined also represents the entire spectrum of advocates of AGW theory, and therefore also includes the so-called "Lukewarmer", who believe that a human contribution to the climate is possible, but reject disaster scenarios for future climate development.

For the predicted global catastrophe scenarios of our future climate development, only a "consensus" of significantly less than a third would remain. And if you then take a very critical look at this background knowledge, the Consensus Project even shows the restriction to the subset described correctly. There it says behind a huge "97% ..." in small print (with own emphasis),

*„... of published climate papers **with a position on human-caused global warming** agree: GLOBAL WARMING IS HAPPENING – AND WE ARE THE CAUSE"*,

so: "97% of the published climate articles with a position on man-made global warming agree: Global warming is happening - and we are the reason". When considering all of Cook et al. evaluated scientific climate publications, the result looks very different:

- A two-thirds majority of the investigated scientific climate work apparently does not make any socio-political statements about AGW.

- Climate realists are only noticeable with about 1% of all examined publications due to their socio-political views against AGW.

- The protagonists of AGW, on the other hand, are less reluctant to make socio-political statements in scientific publications with almost a third of all publications examined.

Result: The ominous and much-quoted consensus for a 97 percent acceptance of the AGW theory does not exist in climate research. And with that, the anti-scientific demand for an "end of the climate discussion" is morally and arithmetically completely on the offside. In the Cook et al. study, however, there is clear evidence that it is essentially the protagonists of a climate catastrophe who contribute sociopolitical positions to scientific work. Finally, in the present study, a comparison of opposed sociopolitical positions in a subjectively selected subset is made the standard for an alleged consensus in all climate sciences.

As a positive result of this study, it can nevertheless be said that in climate research a "silent" two-thirds majority with their scientific work still stays out of the socio-political discussion about the climate catastrophe. In the end, the sociopolitical opinion of a one-third minority is sold to the public as a scientific 97% consensus of the majority in climate research.

Against the background of an "one-third truth" for man-made climate change, it is very strange that the so-called "climate deniers" are always lumped together with deniers of all kinds by the minions of the climate catastrophe. However, it becomes completely incomprehensible if, in an open scientific discussion about the foundations of a feared global climate change, so-called "climate deniers" and "Lukewarmer" are equally opposed to a "climate of hatred" (KalteSonne Article on February 3, 2015), and not just in the UK and the US. For example, in a brochure published by the German Federal Environment Agency in 2013, the critics of man-made climate change were branded as unsuspecting, whereupon the WELT headlined, "*An authority declares the climate debate over*".

Aftermath

On August 16, 2018, ARD-MONITOR ran a program entitled "Climate change and summer heat: the opponents are mobilizing".

The MONITOR forum then had a discussion with co-author Bärbel Winkler from Cook at al. (2013), which subsequently referred to a "parallel logic" and prompted the author to write the following article.

Blog article (reduced): *„Die „parallele Logik" für eine Dekarboni-
sierung der Welt" (U. Weber)*

Published in German at KalteSonne on September 4, 2018:
http://diekaltesonne.de/die-%E2%80%9Eparallele-logik-fur-eine-
dekarbonisierung-der-welt/

Also published at EIKE on September 6, 2018: https://www.eike-klima-
energie.eu/2018/09/06/die-parallele-logik-fuer-eine-dekarbonisierung-der-welt/

The "parallel logic" for decarbonization of the world

During the course of this discussion at the MONITOR forum, the
procedure in the study "Quantifying the consensus on anthropogenic global warming in the scientific literature" by Cook at al.
(2013) derived a "parallel logic" for a 97% consensus for allegedly
man-made climate change (Anthropogenic Global Warming =
AGW). This "parallel logic" was then exposed as the little hat trick
with which the understanding-defining reference variable (2) is
removed from the chained statement (1) -> (2) -> (3), in order to
be then spread to the public as a "parallel truth" (1) -> (3):

(1) First, a group is examined as a 100% total amount (1). The
    underlying question then results in several subsets, each
    with a uniform specificity, which in total add up to this
    100%.

(2) Then any subset (2) with a specific characteristic is out-
    sourced (1) -> (2) and subjected to a closer examination as
    a new 100% basis. Their percentage subsets then relate
    only to the outsourced subset (2).

(3) In the end, a statement (3) is derived for one of these sub-subsets and a percentage from the 100% subsets base (2) is provided, with the original reference (1) -> (2) -> (3) is crucial for the veracity of this statement.

(4) The statement derived from (1) -> (2) -> (3) therefore only remains scientifically valid as long as the "re-labeling" in point (2) is not lost or suppressed.

(5) And now the little hat trick of "parallel logic" comes into play: Sooner or later the statement (1) -> (2) -> (3) becomes directly related to the original total from (1) without the mandatory limitation from (2) and leads to a completely new "parallel truth" (1) - > (3).

Let us now compare this "parallel logic" with the result of Cook et al. (2013):

(1) The summaries of 11,944 scientific papers from the peer-reviewed scientific literature were examined for statements on AGW.

(2) 66.4 percent of the summaries made no statements about AGW and the remaining 32.6 percent with a position on AGW were examined in more detail.

(3) Of these 32.6% with statements about AGW, 97.1% confirm the AGW theory.

The result of the study by Cook et al. (2013) but publicly distributed as a 97% agreement from the total amount of climate articles examined there and as evidence of a scientific 97% AGW consensus, for example by Cook et al. (2016) themselves, they include all publishing climate scientists in their core statement.

The difference between propaganda and science using the example of Cook et al. (2013) amounts exactly 65.4%, the proof below using conventional logic:

„Parallel Logic" [%]: **(1)->(3)** with 97,1% agreement without reference **(2)**

Conv. Logic [%]: (1)->(2)->(3) with 97,1% agreement from 32,6% (1) = **31,7%**

[97,1% from Cook @subset(2)] – [31,7% von Cook @total(1)] = **65,4%**

Mathematical analysis has no "parallel logic" as well as science cannot ascribe a "higher truth" to itself. Rather, historically, the majority of science has always come to terms with the social system in which it has operated. The distinction between truth and ideology is therefore up to each of us ourselves. Consequently, you can either think for yourself or you have to believe everything that is told to you.

Note: "If you don't know anything, you must believe everything"

(Marie von Ebner-Eschenbach)

Note (2019): The number of research papers examined by Cook at al. (2013) is given in the first blog-article (2015) as 12,465 and in the second blog-article (2018) as 11,944. The editor, Environmental Research Letters, said: *"Corrections were made to this article on 31 May 2013. A data file was added to the supplementary data. Further corrections were made on 30 October 2013. A link to further supporting data was added"*. The author cannot rule out to have referred to an early version of Cook et al. (2013) in the 2015 blog-article, which does not change the statements of either blog-article.

A Global CO_2 Budget would be Continuously Renewable

Blog article: *„Prozentrechnung müsste man können: Das en(t)liche CO_2-Budget" (U. Weber)*

Published in German at KalteSonne on April 29, 2017:
http://diekaltesonne.de/prozentrechnung-musste-man-konnen-das-entliche-co2-budget/

On April 11, 2017, an article with the title "*Can we stop global warming in good time?*" was published on the internet blog "Klimalounge". With the statement that a feared rise in temperature of 1.5 to 2 degrees would only allow a global CO_2 budget of 150 to 1050 gigatons (Gt), and then exiting scenarios for the egress from carbon-based fossil fuels were presented. There it is claimed that the temperature level at which global warming would later come to a halt would, to a good approximation, be proportional to the cumulative CO_2 emissions and to stop global warming, global zero emissions for CO_2 would have to be reached before 2050.

But even if you believe in man-made climate change through the use of fossil fuels, you shouldn't get scared. From a scientific point of view, it certainly does no harm to take a closer look at the IPCC graphic shown below on which the reasoning is based and compare it with additional facts:

The climate impact of CO_2 is usually given as "climate sensitivity" in degrees per doubling. The IPCC specifies a range from 1.5 to 4.5 [°C / 2xCO_2]. The original pre-industrial atmospheric CO_2 content is said to have been 280 ppm. By 2015, humans had brought about 1400 Gt CO_2 additionally into the atmosphere from the use of fossil fuels and thus increased the CO_2 content of the atmosphere to 400 ppm. Here is the IPCC image from the Klima Lounge article dated April 11, 2017:

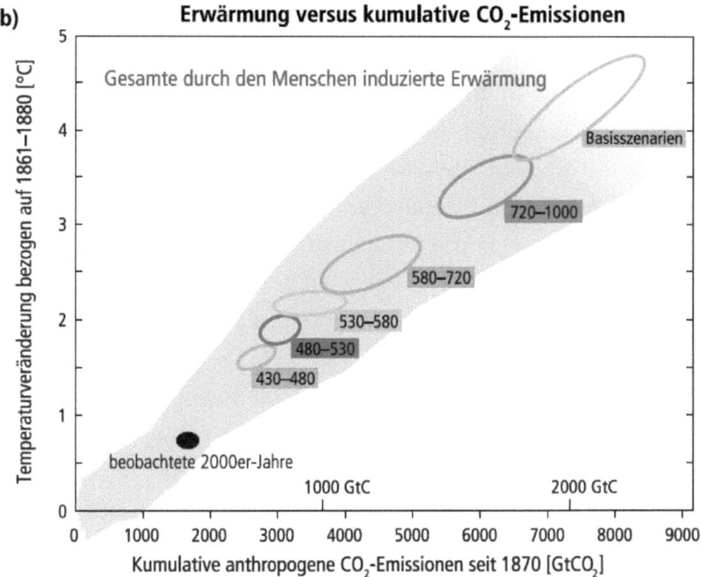

Cumulative anthropogene CO$_2$-Emissions since 1870 (Gigatons)

[Inserted source (2019): IPCC, Fig.SPM.5b from AR5 (2014) SYR]

The text for this figure states there, quote: *"Relationship between cumulative CO$_2$ emissions and global warming. The numbers on the "bubbles" indicate the CO$_2$ concentration in the atmosphere achieved in the different scenarios. The temperature indicated on the vertical axis applies at the point in time when the emission quantity indicated on the horizontal axis is reached. That means: the following further heating solely due to the thermal inertia in* *the system is not yet taken into account here. Source: IPCC synthesis report (2014). "*

The statements about the remaining global CO_2 budget are apparently based on the bubble value from the IPCC graphic above (480-530 ppm = 3,000 Gt CO_2 = 1.75-2.0 dT°C). Let's compare these values with the additional facts listed above: The pre-industrial CO_2 content in our atmosphere should have been 280 ppm or 0.028%. For the period between 1900 and 2015, the anthropogenic CO_2 input totaled around 1400 gigatons (Gt) and had led to an increase in the atmospheric CO_2 content by 0.012% to 0.040% or 400 ppm. The relationship between anthropogenic CO_2 emissions and atmospheric CO_2 content is as follows:

(1) $\qquad$ X Gt CO_2 = 280 ppm

with X = "natural" atmospheric amount of CO_2 [Gt CO_2]

(2) $\qquad$ X Gt CO_2 + 1.400 Gt CO_2 = 400 ppm

(3) = (2) − (1) $\qquad$ 1.400 Gt CO_2 = 120 ppm

The original total amount of atmospheric CO2 "X [Gt CO_2]" is then obtained from lines (1) and (3) using a simple rule of three calculation:

X Gt CO_2 = 280 ppm x 1.400 Gt CO_2 / 120 ppm = 3.200 Gt CO_2

We can carefully take a mean CO_2 concentration of 510 ppm from the IPCC bubble value with (480-530 ppm = 3,000 Gt CO_2 = 1.75-2.0 dT °C) for a maximum anthropogenic temperature rise below 2 degrees. This 510 ppm then corresponds to almost 6,000 Gt CO_2. Deducting the natural, atmospheric amount of CO_2, this results in an original global emissions budget of 2,800 Gt CO_2 for a temperature increase below 2 degrees, which is even 200 Gt CO2 smaller than stated in the IPCC bubble. Of this originally available global CO_2 budget of 2,800 Gt CO_2, 1,400 Gt CO_2 would have been "used up". Incidentally, this value also agrees very well with the information provided by the Federal Agency for Civic Education, according to which the concentration of CO_2 is said to have increased by around 40 percent since industrialization began.

According to the estimate carried out here, an increase in the pre-industrial atmospheric CO_2 content to 510 ppm would therefore require a further 1,400 Gt CO_2 (= 2,800 Gt CO_2– 1,400 Gt CO_2) in order to finally increase the overall atmospheric temperature according to the IPCC graphic shown above about 1.75-2.0 ° C. With a worldwide annual CO_2 emissions of constantly 30 gigatons, it would take about 45 years from 2015 to an allegedly anthropogenic temperature rise of just under 2 degrees Celsius, i.e. until 2060.

However, the statement about a cumulative effect of CO_2 for measuring the available CO_2 budget in the climate lounge article cited is only correct to the extent that this CO_2 is still in the atmosphere. The anthropogenic CO_2 has a residence time of only about 120 years in our atmosphere (German Federal Environment Agency under the keyword "carbon dioxide").

The global CO_2 budget is therefore not cumulative at all, but continuously "renewable"!

Note (2019): The information from the German Federal Environment Agency linked in the blog-article has increased eightfold without reason from "120 years" to "up to 1000 years". You will find the CO_2 retention period of 120 years on the Internet if you search for "CO_2 retention period" or "CO_2 budget" - but not in the IPCC 1.5 ° C report, which in turn is based on "cumulative CO_2 emissions"...

With this residence time of around 120 years for anthropogenic CO_2 in our atmosphere, the ongoing global CO_2 budget for anthropogenic CO_2 emissions is around 2,800 Gt CO_2 per 120 years.

Consequently, a supposedly man-made temperature rise is likely to remain below 2 degrees. And we actually have time until 2060 to limit the anthropogenic CO_2 emissions to 23 gigatons annually (= 2,800 Gt CO_2 / 120 years) and thus to permanently secure the ominous 2-degree target. There can be no question of zero emissions from 2050 even if you really want to believe in man-made climate change through CO_2 emissions.

A annual global CO_2 budget of around 23 gigatons of CO_2 would rather keep the feared man-made climate warming permanently below 2 degrees.

The decarbonization of the world has again proven to be a completely unnecessary self-mortification of humanity. And based on this religious agenda, the western world now wants to voluntarily destroy its carbon-based livelihood and destroy our resulting standard of living. The only question that remains is whether the consenting silence of a social majority in the western industrialized nations threatened by global decarbonisation is based on over-saturated disinterest, a stunted individual ability to criticize or inadequate knowledge in the percentage calculation...

CO_2 is not the Natural Global Climate Driver

Scientific paper: WEBER, U.O. (2016): *„About the Natural Climate Driver"* published in the „Mitteilungen der Deutschen Geophysikalischen Gesellschaft" 2/2016: 9-11

Abstract

Future climate computer models do not separate between natural and anthropogenic climate forcing. Sporadically, scientific papers try to introduce carbon dioxide (CO_2) to be the dominant natural climate driver. An analysis from paleoclimate temperature proxies with the attributed maximum climate sensitivity of carbon dioxide leads to the result that CO_2 could not account for the natural (paleo-) climate variability. A further IPCC-conform calculation with historical data for temperature and atmospheric CO_2-contents determines the climate sensitivity of CO_2 about its published minimum value resulting in the finding that CO_2 could not be the dominant natural climate driver.

Climate Forcing

More than thirty years of intensive climate research did not lead to a separation between natural and anthropogenic climate forcing in future-climate models yet. In geosciences, paleoclimate changes are attributed to the orbital variations of the earth (Milankovic´ cycles) beginning with KÖPPEN & WEGENER (1924). But the small portion of absolute solar energy variance during these orbital cycles could not account itself for the amount of such paleoclimate variations while there is an imperative correlation between each other in the frequency domain. Only secondary forcing like the ice-albedo effect could account arithmetically for the magnitude of global temperature change during past ice-ages (WEBER 2o15). Sequence stratigraphy, which has been established during the past decades, considers the orbital Milankovic′ cycles as dominant factors and has been proven to be a reliable geological tool for the understanding of sedimentary processes. Sporadically, scientific papers try to introduce CO_2 as the dominant natural climate driver, so recently SHAKUN et al. (2o15), while their own frequency analysis of an "ice-volume CO_2 gain function" confirms the close correlation between paleoclimate and orbital changes. In the following approach, the effective influence of CO_2-forcing on the earth's climate variability will be examined from temperature proxies and measured data.

Contribution of CO_2 to Climate Forcing

The Intergovernmental Panel on Climate Change (IPCC 2o01) describes the climate sensitivity parameter of CO_2 (global mean surface temperature response ΔTs to the radiative forcing ΔF) for the temperature-equivalent of the radiative CO_2-forcing as:

(1) $\Delta F = \alpha \times \ln (C/C_o)$.

Note: The function "ln" will be replaced in the formulas (3) to (6) by "log2" or "2^", respectively, because this formula describes the effect of doubling (of the atmospheric CO_2-contents).

With α = 5.35 W/m² and $\Delta Ts/\Delta F$ = λ = 0.5 °C/W/m² from IPCC (2oo1) one receives:

(2) $\Delta Ts = \lambda \times \alpha \times \ln (C/C_o)$ [°C].

From the IPCC-formula (2), we may consider the product ($\lambda \times \alpha$) (°C) to represent the climate sensitivity of CO_2 for the temperature effect of changes in its atmospheric contents. The temperature effect of the arbitrary atmospheric CO_2-contents C on the global average near-surface temperature at the pre-industrial atmospheric CO_2-contents Co of 28o ppm could then be given by the equation:

(3) $\Delta T = CS_{CO2} \times \log_2 (C/C_o)$

        with CS_{CO2}: climate sensitivity of CO_2 (°C)
        and CO_2-contents C and C_o (ppm).

Estimate for a Sole Paleoclimate Forcing from CO_2

Formula (3) could be resolved for the arbitrary CO_2-contents and then recalculated with the Vostok temperature proxy data for a sole climate forcing from the theoretical atmospheric paleo-CO_2-contents C:

(4) $C = 2^{\wedge}(\Delta T/CS_{CO2}) \times C_o$

            with ΔT: Vostok temperature proxies (°C),
            $CSCO_2$ = 4.5 °C:climate sensitivity of CO_2,
            CO_2-contents C and Co (ppm), and Co = 28o ppm.

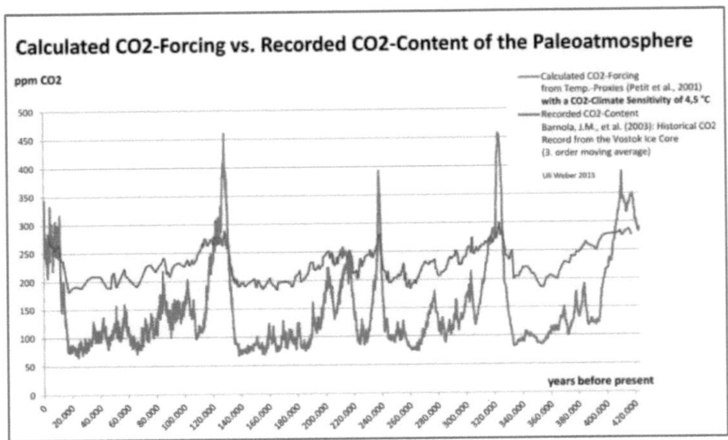

Fig. 1: Calculated CO$_2$-forcing from temperature proxies (blue [=Curve runs between 65 and 46o ppm]*) and measured aerial CO$_2$-portion from the Vostok icecores (red* [=curve between 18o and 3oo ppm]*)*

In Figure 1, the climate sensitivity of CO$_2$ at the maximum IPCC-value of 4.5 °C for a doubling of the atmospheric CO$_2$-contents (IPCC 2o13: 1.5 °C to 4.5 °C with high confidence) has been applied to the Vostok temperature proxies from PETIT et al. (2oo1), starting with the pre-industrial CO$_2$-contents of 28o ppm at o °C temperature proxy. The calculated paleo-atmospheric CO$_2$-contents is indicated by the blue graph and represents the theoretical amount of a sole CO$_2$-paleoclimate forcing for the given temperature proxy data. The red graph shows the measured CO$_2$-contents of the paleo-atmosphere from the Vostok ice cores analysed by BARNOLA et al. (2oo3). It becomes obvious that a theoretical CO$_2$-forcing derived from the Vostok temperature proxies requires much wider spreading CO$_2$-contents of the paleo-atmosphere (approx. 65 to 460 ppm) than indicated by the real measured atmospheric CO_2-values (approx. 180 to 300 ppm). Reasonable smaller values towards the minimum IPCC-value of 1.5 °C for the climate sensitivity of CO_2 would result in an even larger calculated CO_2-variance. Reversely calculated, a CO_2-forcing at a climate sensitivity of 4.5 °C applied to the measured paleo-CO_2-contents would lead to a temperature variance between −2.87 °C (at 180 ppm) and +0.45 °C (at 300 ppm) for the past 420,000 years, compared to the Vostok minimum/maximum-temperature proxies ranging from −9.39 to +3.23 °C. A resolution of the above given IPCCconform equation (3) regarding the CO_2-climate sensitivity CS_{CO2} yields:

(5) $CS_{CO2} = \Delta T/ \log_2(C/C_o)$ (°C).

This formula allows a recalculation of the CO_2-climate sensitivity for a sole CO_2-forcing of the paleo-atmosphere from the Vostok temperature proxies. Due to uncertainties of the combination of the data pairs (Vostok temperature proxy at paleo-atmospheric CO_2-contents) and divisions by small numbers, the variation of the results is quite large for the above given minimum/maximum-values (180 ppm at −9.39 °C = 14.7 °C and 300 ppm at +3.23 °C = 32.5 °C, respectively), while the average of the whole dataset results in a mean climate sensitivity of 16.7 °C. Thus, a CO_2-climate sensitivity of about 16 °C would be required to meet the Vostok temperature proxies for a sole climate forcing from the paleo-atmospheric CO_2-contents. Such climate sensitivity of CO_2 is far off the range of confidence given by the IPCC (2013: 1.5 to 4.5 °C with high confidence) proving that CO_2 could not be the dominant natural climate driver.

Estimate for the Anthropogenic Climate Forcing

According to various reports of the IPCC, CO_2 is the predominant anthropogenic climate driver. In the period between the years 188o and 2o12 the atmospheric CO_2-contents rose from pre-industrial value of 28o to 394 ppm (NOAA) from anthropogenic sources. According to formula (3), an approximation for the global temperaturerise from the beginning of industrialisation as a sole anthropogenic CO_2-forcing can be calculated. The minimum IPCC-value of 1.5 °C for the climate sensitivity of CO_2 (IPCC 2o13: 1.5 to 4.5 °C) for the given data (28o ppm CO_2 at o °C and 394 ppm CO_2 at ΔT) results in a global temperature rise of o.74 °C:

(6) $\Delta T = 1.5 \times \log_2 (C = 394 \text{ ppm}/C_o = 28o \text{ ppm}) = o.74 \text{ °C}$

(calculated temperature-rise since 188o).

For the same period, IPCC (2o14) reports a rise of the globally averaged surface temperature of o.85 °C. It has to be clearly stated here that the reported temperature rise must include the entire anthropogenic forcing and their atmospheric feedback and additionally the effects of aerosols and grime cooling, which are mainly bound to the generation of CO_2. The difference of 13% between the calculated temperature rise from a sole CO_2-forcing and the effective temperature rise given by the IPCC (2o14) include the climate forcing of minor anthropogenic greenhouse gasses (IPCC 2oo1) and its atmospheric feedback. The above calculated result confirms in a first approach the magnitude of the effective CO_2-climate sensitivity to account for the reported IPCC minimum-value.

Natural and Anthropogenic Climate Forcing

Paleoclimate proxy data prove that natural climate forcing from solar radiation has never been constant through the geological and historical past. And because natural climate forcing persists through geological times until today and will continue in future, the temperature change since industrialisation could not be a sole anthropogenic effect alone. Consequently, for any given time t after industrialisation both, natural and anthropogenic forcing, must have contributed to the global climate $C_{WW}(t)$:

(7) $\qquad C_{WW}(t) = F\,(F_N(t) + \Delta F_A(t))$

$\qquad\qquad$ with $F_N(t)$ natural climate forcing
$\qquad\qquad$ and $\Delta F_A(t)$ anthropogenic climate forcing (= o at t < year 185o)

Conclusively, even if mankind could restrain from future active climate forcing, such global economic effort could never lead to a constant natural climate on earth nor could it prevent from future climate changes. The above given estimate for the effective climate sensitivity of CO_2 derived from formula (6) still contains the change of natural climate forcing since the beginning of industrialisation. The final result for the objective climate forcing capabilities of CO_2 is then given by the lower limit from the IPCC (2o13) diminished by the factual change of natural climate forcing between the years 188o and 2o12.

Conclusion

From the presented approach to the climate forcing capabilities of carbon dioxide (CO_2) follows conclusively that CO_2 could not be the primary natural climate driver. Even a maximum climate sensitivity of 4.5 °C according to the IPCC could not account for paleoclimate changes. The reported amount of rise in temperature and atmospheric CO_2-contents since the beginning of industrialisation leads to an effective climate sensitivity of CO_2 at about the minimum value attributed by the IPCC (1.5 °C). The change of natural climate forcing since industrialisation is still included in this approach and diminishes the climate sensitivity of CO_2 with its factual share below the published minimum. Climate sciences are seriously addressed to focus research on the natural climate forcing mechanism. Only a quantitative separation between natural and anthropogenic climate forcing would allow credible climate models in future.

Acknowledgement

The author thanks the DGG editorial team for improvements in the manuscript and their kind support in adapting it to the DGG publication format.

References

• BARNOLA, J.-M., RAYNAUD, D., LORIUS, C. & BARKOV, N.I. (2003): Historical CO2 record from the Vostok ice core. – In: Trends: A Compendium of Data on Global Change. Oak Ridge (Carbon Dioxide Information Analysis Center, Oak Ridge National Laboratory, U.S. Department of Energy).

• IPCC (2001): Radiative Forcing of Climate Change 2001 – The Scientific Basis. – <www.grida.no/climate/ipcc_tar/wg1/pdf/TAR-06.pdf>: 354, Tab. 6.2.

• IPCC (2013): Climate Change 2013 – The Physical Science Basis – Summary for Policymakers. – <
www.ipcc.ch/pdf/assessmentreport/ar5/wg1/WGIAR5_SPM_brochure_en.pdf >: Climate Sensitivity of CO2 – D.2 Quantification of Climate System Responses.

• IPCC (2014): Climate Change 2014 – Synthesis Report – Summary for Policymakers. – < www.ipcc.ch/pdf/assessment-report/ar5/syr/AR5_SYR_FINAL_SPM.pdf >: SPM 1.1 Observed changes in the climate system.

• KÖPPEN, W. & WEGENER, A. (1924): Die Klimate der geologischen Vorzeit. – Berlin (Bornträger).

• NOAA Global Greenhouse Gas Reference Network: Trends in Atmospheric Carbon Dioxide Mauna Loa Recent Monthly Average CO2. – <
www.esrl.noaa.gov/ gmd/ccgg/trends/ >.
Last access:2015-12-11.

• PETIT, J.R., et al. (2001): Vostok Ice Core Data for 420,000 Years. – IGBP PAGES/World Data Center for Paleoclimatology. Data Contribution Series #2001-076, NOAA/NGDC Paleoclimatology Program; Boulder.

• SHAKUN, J.D., CLARK, P.U., HE, F., LIFTON, N.A., LIU, Z. & OTTOBLIESNER, B.L. (2015): Regional and global forcing of glacier retreat during the last deglaciation. – Nature Communications, 6: 8059, doi: 10.1038/ncomms9059.

• WEBER, U. (2015): An albedo approach to paleoclimate cycles. – Mitteilungen der Deutschen Geophysikalischen Gesellschaft, 3/2015: 18-22.

A Natural Albedo Forcing Explains the Global Climate Genesis

Scientific paper: WEBER, U.O. (2015): *„An Albedo Approach to Paleoclimate Cycles"* published in the „Mitteilungen der Deutschen Geophysikalischen Gesellschaft" 3/2015: 18-22

Abstract

The contribution of the solar irradiation to prominent changes of the earth's average temperature is strongly questioned by global climate computer models although there is a close correlation between the frequencies of paleoclimate changes from temperature proxy series and the variability of the earth's orbital Milankovic´ cycles. The earth's spherical albedo could account for paleoclimate changes if it would be modulated by variations of the natural insolation at high latitudes through the changes of the earth's orbital parameters. The recent albedo of a = 0.3016 represents the actual areal distribution of land masses, oceans, green lands, deserts, mountains, and ice-sheets as well as an averaged cloud cover. This paper shows that with variations of the earth's albedo between 0.2801 and 0.3640 the variability of the Vostok temperature proxies could be met.

Paleoclimate Changes

The resampled proxy-temperatures derived from the Vostok ice-cores (PETIT et al. 2001) presented in Figure 1 show an oscillation between +3.23 and -9.39 °Celsius against the temperature of 0°C set at year o from sampling. These temperature proxy series indicate no "natural" constant global temperature level at any time period. The curve-trend shows a present short-time climate optimum and considerably lower global temperatures as present for most of the past 42o,ooo years (yr).

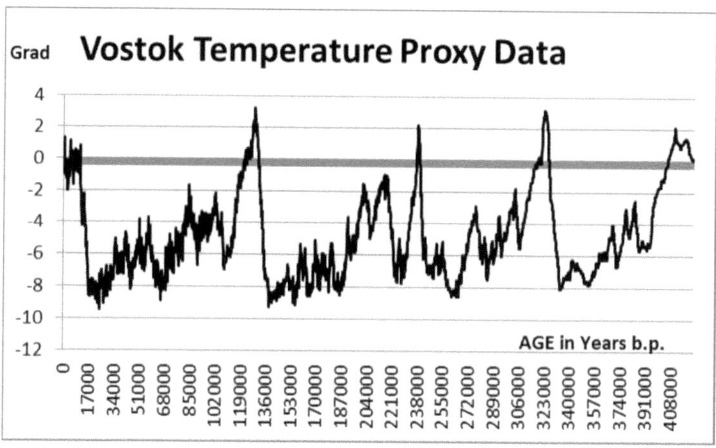

Fig. 1: Resampled proxy-temperatures derived from the Vostok ice-core data (PETIT et al. 2001); Grad in °C

On a first sight the Vostok temperature proxies may allow to identify three different phases of paleoclimate: The Vostok temperature proxies show an actual climate optimum for the past 12,ooo yr about the recent global mean temperature. From the past four ice-age cycles could be derived that the duration of such an optimum averages about 1o,ooo to 15,ooo yr ending with a drop to a mean temperature between -2 und -6 °C for a 5o,ooo years period. This era is finally followed by another period of about 5o,ooo yr with a drop to -4 to -8 °C against the actual temperature. Such complete ice-age cycle lasts then about 115.ooo yr while the average temperature of the whole Vostok temperature proxy series represents a mean value of about -4.5 °C.

Thesis: *Setting the Vostok 0 °C temperature to the actual global mean near-surface temperature of the earth (NST) at 14.83 °C we receive a variability of absolute Vostok proxytemperatures between 5.44 and 18.06 °C around a Vostok median temperature (VMT) of 11.75 °C. As a first paleoclimate approach, the Vostok proxy-temperatures (VPT) have alternated through the past 420,000 yr (t) around the actual global mean temperature (NST) at*

$$V_{PT}(t) = NST +3{,}23\,°C\,/\,{-9{,}39}\,°C.$$

The question in dispute is the source of climate forcing for such temperature variability throughout the past.

Possible Primary Forcing of Paleoclimate Variations

Four types of climate forcing could be identified to possibly account for changes of the global mean temperature through paleoclimate history:

• Cyclical changes of the natural solar irradiation itself,

• Cyclical changes of the solar irradiation through the earth's orbital variations,

• Primary terrestrial energy sources,

• The natural greenhouse effect (NGE) describing the climate forcing capabilities of so-called climate active gases, predominantly water, carbon dioxide, and methane, fed by infrared back-radiation of the earth's surface and global circulations.

The natural solar flux shows only a variability of about o.1 %, i.e. around 1.4 W/m². SCHWARZ (no date) calculates average annual changes of the solar flux caused by the geometry of the earth's orbital variations of about o.1 % as well. Obviously, any combination of the variance of the sun's strength and orbital-caused variations of the incoming solar insolation would not exceed o.2 % of the total solar flux. Consequently, neither the variability of the natural solar cycles nor the orbital Milankovic´ cycles alone, nor any combination of both could account directly for paleoclimate changes. In contradiction there persists an excellent spectral correlation between orbital variations and climate proxy series. Terrestrial energy sources, except volcano eruptions with a multi decade climate influence, count with but a fraction of the variability of the primary solar forcing while their contribution is either constant or accidental. The accidental fraction declines with an increasing period of observation ending up in an averaged constant contribution. There is also no evidence that the greenhouse effect, i.e. absorption of low frequency IR back radiation mainly by the "climate gases" water vapor, carbon dioxide, and methane could count for variations of the global mean temperature in the magnitude which is given by paleoclimate temperature proxies. The climate sensitivity of CO_2 is estimated by different sources between 2 and 4.5 °C per doubling of its atmospheric content. In a rough estimate for the Vostok mean temperature (11.75 °C) at a pre-industrial CO_2 content of 28o ppm this atmospheric CO_2 content must then have varied between 14o ppm and 56o ppm to account for the past variations of the paleoclimate temperature proxies. No such magnitude of variance of the atmospheric CO_2

content has ever been reported for the past two million years. Furthermore, there is not any active mechanism known to account directly for the relation between the concentration of greenhouse gases and the temperature proxy series nor is there any fixed relation given between their concentration and the natural variations of the global mean temperature through paleoclimate periods. And finally, there is high evidence from temperature proxy analyses that a rise of the atmospheric CO_2 content always follows a rise of the mean global temperature. SHAKUN et al. (2015) show the ratio of spectral power of an ice-volume reconstruction to the ice-core CO_2 record over the past 800 kyr after normalizing each series to mean zero, unit variance (Fig. 2).

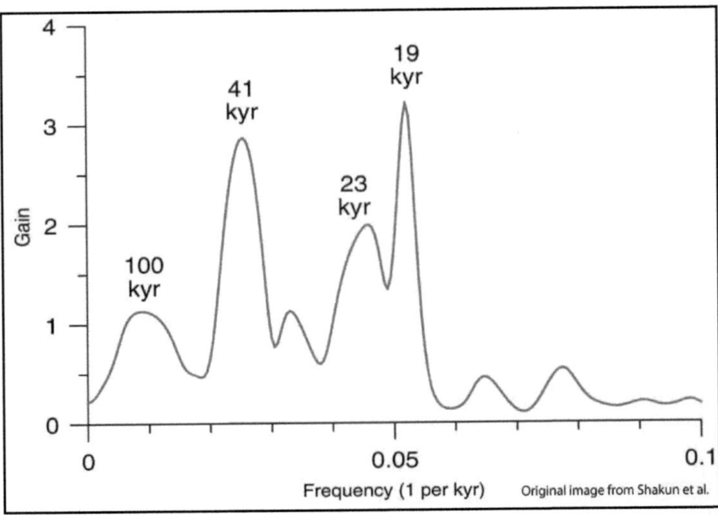

Fig. 2: The ice-volume CO2 gain function from SHAKUN et al. (2015, licensed under the Creative Commons Attribution 4.0 International License)

Figure 2 clearly indicates a strong dependence of the changes in atmospheric CO2 content and the ice volume from the periods of eccentricity (1oo kyr), obliquity (41 kyr) and precession (19 and 23 kyr) of the earth's orbital parameters (Milankovic´cycles). Neither the ice volume nor CO_2 are primary energy forcings, but secondary effects on that. Both demand consequently variations of their own extent to have an effect on the global temperatures. Consequently, if there is no active climate forcing responsible for the earth's climate variability through geological times, this variability must then be caused by a prominent secondary effect. This paper shows that a modulation of the natural solar insolation by changes in the earth's albedo could count for the variations of the earth's average temperature through geological times.

Relation between Temperature and the Natural Solar Insolation

The energy forcing from the sun counts at 1,367 W/m² with an average of 342 W/m² for the overall surface of the earth at a global mean temperature of 14.83 °C. DOUGLAS & CLADER (2oo2) give the climate sensitivity for variations of the solar forcing with

$$\Delta T = k \cdot \Delta F$$

with T: Temperature (°C) and *F*: Forcing (W/m²).

They have calculated a sensitivity coefficient k from measured short-time data series of temperatures and solar irradiation at

$$\Delta T / \Delta F = k = 0.11 \pm 0.02$$

sensitivity coefficient in (°C / Wm-2)

(DOUGLAS & CLADER 2oo2).

Consequently, to meet the Vostok proxy temperature (TVP) variations between +3.23 and -9.39 °C with respect to the actual global mean temperature (NST) of 14.83 °C the past climate forcing of the earth would represent variations of

$$\Delta F_{V@NST} = +29.36 \text{ W/m}^2 \text{ and } -85.36 \text{ W/m}^2$$

around the actual global mean of 14.83 °C.

Variations in climate forcing of about ±57 W/m² are needed to meet the variations of the Vostok temperature proxies around the Vostok median temperature of 11.75 °C. Consequently, the temperature changes within the paleoclimate cycles demand energy variations of about 1oo-fold the variability of the natural solar cycles or the variability of the net solar flux caused by orbital variations of the earth, respectively.

LASKAR et al. (1993: Fig. 5) come with their approach to calculate the change of solar flux for a fixed geographical location at 65 °N and 12o °E in the northern hemisphere through the orbital variations of the earth to changes in insolation up to ±5o W/m² with a periodic beat frequency around 25 kyr for the past 1 million years. The variability of solar forcing in higher latitudes given by LASKAR et al. (1993) corresponds well with the here calculated variation of about 115 W/m² derived from the Vostok temperature proxies.

Modulation of the Solar Insulation to meet the Vostok Proxy Temperatures

A direct relation between a primary forcing and the changes of the earth's average temperature through geological times could not be established here. Consequently, a prominent secondary effect must be modulating the solar insulation to meet the Vostok temperature proxies. The only possible secondary forcing is the low-frequency back-radiation of the earth that empowers the global circulations. The amount of the effective solar climate forcing is controlled by the earth's albedo.

With the relation between the climate sensitivity λ and the climate sensitivity coefficient k the earth's albedo is given by DOUGLAS & CLADER (2oo2) at

$$\lambda = \frac{4}{1-a} * k \qquad \text{we receive for the earth's albedo}$$

$$a = 1 - \frac{4 * k}{\lambda} \qquad = 0.3016 \text{ with } k = 0.11 \text{ °C/Wm}^{-2}$$

$$\text{und } \lambda = 0.63 \text{ °C/(Wm}^{-2}).$$

Question is how much the earth's albedo must change to cause the needed variability in solar energy forcing to meet the Vostok temperature proxies.

From the solar insolation of 1,367 W/m² solar forcing at top of the atmosphere the albedo value derived from DOUGLAS & CLADER (2oo2) with $a = 0.3016$ represents then an average reflected/refracted energy portion of 412.29 W/m² which does not contribute to climate forcing. In a first approach the minimum and maximum albedo values needed to meet the Vostok temperature proxies around the actual global near-surface temperature NST of 14.83 °C have been calculated. Base for this calculation is the reflected portion of the solar insolation of 412.29 W/m² at the recent global albedo of 0.3016 resulting in a flux change of 13.67 W/m² per percent albedo. The variability of the albedo is then given by

$$F@a_{min} = (412.29 - 29.36) \text{ W/m}^2 = 382.93 \text{ W/m}^2 \text{ mit } a_{min} = 0.2801$$

$$F@a_{max} = (412.29 + 85.36) \text{ W/m}^2 = 497.65 \text{ W/m}^2 \text{ mit } a_{max} = 0.3640$$

From this approximation the earth's albedo may have varied in the past 420,000 yr between $a$min = 0.2801 and $a$max = 0.3640 as shown in Figure 3 to meet the Vostok temperature proxies presented in Figure 1. The portion of cloud albedo (aerosol and Svensmark effect) over paleoclimate cycles cannot be estimated separately and is already included included in these values.

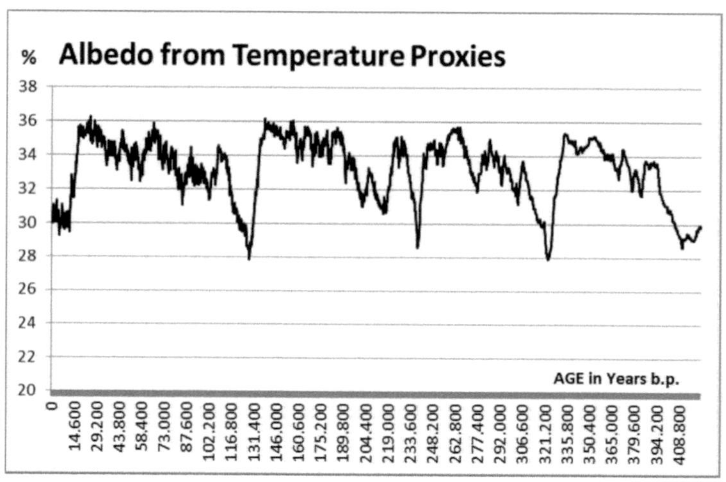

Fig. 3: The variations of the earth's albedo derived from the Vostok temperature proxies (PETIT et al. 2001)

The albedo of ice and snow can rise up to 9o %, but the solar energy density at ground level decreases with the cosine of latitude. At 6o° latitude the cosine is o.5 expressing that the earth's surface receives only half the amount of solar energy per square meter compared to the equator. The areal extent on the earth's surface needed for such secondary ice albedo forcing would consequently mount up between double and threefold the calculated percentage of the albedo variations between -2.15 and +6.24 %. As a rough estimate for the areal extent of such ice albedo effect variations of the earth's snow and ice cover between about -6.5 and +19 % of the earth's surface are needed to meet the Vostok temperature proxies.

Glacial "Runaway Warming"

The Vostok temperature proxies demonstrate that the rise to a climate optimum is very fast compared to the drop of temperatures into ice-ages. But variations in orbital parameters, solar insolation, and the geographical distribution of glaciation usually follow continuous functions which use to have no singularities. If we assume that the Vostok average temperature of about -4,5 °C against the present NST may represent an average climate situation through the past paleoclimate cycles then the earth may reside in a permanent ice-age scenario through the past million years as being represented by the actual glaciation of Antarctica and Greenland.

Thesis: *It may be a typical subjective human viewpoint to postulate a "Runaway Glaciation" from a climate temperature optimum period. The paleoclimate changes through the past ice ages may be better understood from the rise of temperature. Such "runaway warming" should be controlled by orbital changes and we just have to flip arguments to understand a permanent ice-age scenario at rising insolation.*

Comparing the global climate through the past million years with orbital parameters and the average insolation at 65 °N in Figure 4 there seems to be a reasonable coincidence of such short-time climate optimum periods with maxima of eccentricity, despite of their absolute value and maximum insolation at 65 °N. Orbital eccentricity causes a difference in solar flux between perihelia and aphelia of 2 % at its minimum and more than 23 % at its maximum value (Wikipedia 2o13); the actual value between perihelia and aphelia accounts for 6.9 %. This biannual disparity of solar flux is equalized over the year by a higher orbital track speed at perihelia compared to aphelia.

Therefore, the variability of the accumulated annual solar flux caused by eccentricity does not exceed o.1 % (SCHWARZ, no date). CLARK et al. (2o12: Fig. 3C) show that the summer insolation at 65 °N peaks in a final stage of glaciation while the winter insolation rises then through the phase of deglaciation.

But neither the winter insolation nor the total annual amount of solar energy could be mandatory for the process of deglaciation from ice-ages.

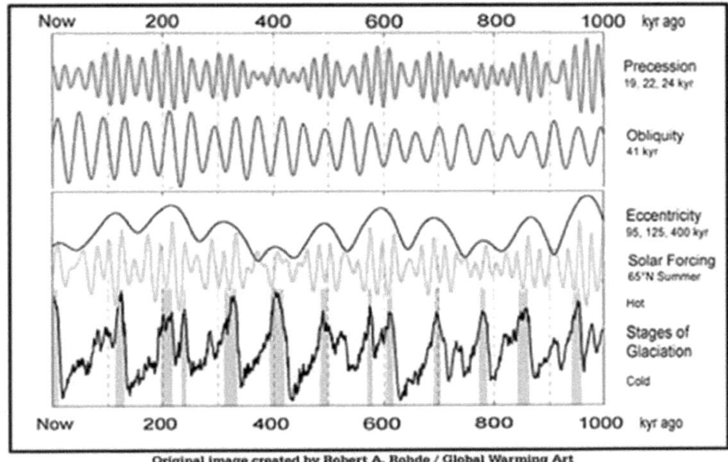

Fig. 4: Diagram of the orbital Milankovi⬚ cycles (from Wikimedia Commons (2013, image created by ROHDE, R.A., Global Warming Art, licensed under the Creative Commons Attribution-Share Alike 3.0 Unported License)

Consequently, a "weak-point" must support the small orbital summer forcing at higher latitudes to cause the global temperature rise into a warm interval. Compared to the stable situation at Antarctica this "weak-point" could only be the Arctic Ocean representing the major area of possible glaciation within the Northern Polar Circle. At ice ages, the frozen Arctic Ocean represents a mirror for solar insolation. Abrupt changes of northern glaciation would be possible if the Arctic Ocean loses its ice-cover through the summer time and becomes a "hot spot". ALLEY & CLARK (1999) show that the North Atlantic Deep Water Flow (NADW) has a great influence on northern deglaciation. The "switch" for deglaciation of the northern hemisphere may then be a maximal insolation at higher latitudes (orbital precession and obliquity) at increasing orbital eccentricity and is represented by the expression:

Global "Runaway Warming" (RW)

@ (Max Insolation @ 65°) + (Eccentricity rising to Max)

At "normal" ice-age situations the spread of glaciation in the northern hemisphere is not limited by energy input from marine circulations nor from material loss from icecalving. Ice volume losses through ice-calving northward into the frozen Arctic Ocean would have no major effect while southward ice drift would be restricted to the North Atlantic and the Bering Strait. The Arctic Ocean is mainly surrounded by land masses at about the Arctic Circle. Consequently, massive further growth of glaciation upon the continental masses of North America and Eurasia is then possible. Obviously, only the Northern glaciation could be modulated through a coincidence of positive peaks in orbital eccentricity because the Southern glaciation at Antarctica persists through the present climate optimum on a continental scale. It seems that a maximum insolation at 65 °N could cause a "runaway warming" resulting in a breakdown of northern glaciation and a global warm interval. SHAKUN et al. (2o15) have calculated temperatures from orbital summer insolation for the western United States at 45 °N for the past 25,ooo yr (Fig. 5). Between the years 2o,ooo and 12,ooo this model shows a remarkable temperature rise of about 6 °C. After this peak the modelled temperatures drop again by about 5 °C through deglaciation until 3,ooo B.C.

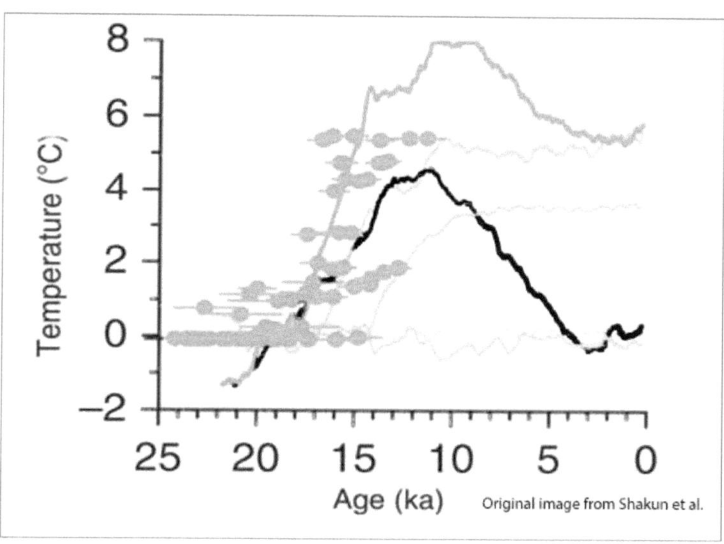

Fig. 5: Modelled temperatures from summer insolation (June-July-August) at 45 °N (bold black line) for the western United States from SHAKUN et al. (2015, licensed under the Creative Commons Attribution 4.0 International License).

Please note: The shown model time series from SHAKUN et al. (2015) is a 500-year moving average and given as an anomaly from 19 kyr. The grey curves are different forcing simulations not used here for argumentation, dots mark normalized moraine positions.

At a first glance this orbital caused temperature trend could not account itself for an ongoing interglacial warm period. But with the here presented model of instable glaciation in the northern hemisphere the drop of orbital solar forcing at the deglaciation begin is no longer a contradiction to further deglaciation. The peaking orbital solar forcing could be understood to push the secondary albedo forcing. The increase of orbital forcing may cause both, a weakening of the ice cover at the Arctic Ocean in summer and a glacier retreat at the Southern extent of Northern glaciation, both resulting in a drop of albedo. This drop as a positive feedback mechanism will then proportionally increase the effective global net solar forcing resulting in increasing global average temperature. When the orbital forcing drops finally again, such secondary albedo forcing seems obviously strong enough to continue deglaciation. On the other hand a new glaciation cycle will start at high latitudes with rising albedo against an increasing solar flux to lower latitudes. This mechanism may explain for the imbalance in rise and fall of temperatures through paleoclimate cycles.

Conclusion

The Milankovic´ cycles are the only time series that could meet the variations of the Vostok temperature proxies in frequency, while the earth's albedo is the only variable that could account for the magnitude of paleoclimate temperature variations. This secondary climate forcing by the albedo is comparable to an electronic amplification circuit, where a small base signal, here the summer insolation between 45 °N and 65 °N, regulates the effective global solar insolation through variations of the earth's albedo.

To meet the Vostok temperature proxies in a simplified approximation the earth's spherical albedo should have varied between the extremes $a$min = o.28o1 and $a$max = o.364o according to the orbital Milankovic´cycles. There is evidence that the true "natural" climate situation of the earth through the past million years is represented by an ice-age scenario with albedo values around 34 %. Depending on variations of the orbital parameters of the earth the ice albedo effect could modulate the natural solar irradiation through growing and declining ice-sheets. If maximum values of insolation at high latitudes from variations in orbital precession and obliquity come together with positive peaks in orbital eccentricity this coincidence could obviously cause a sudden breakdown of glaciation in the Northern hemisphere by a "runaway arctic warming" to meet the Vostok maximum temperature proxies.

Acknowledgement: The author thanks Dipl.-Geophys. Birger Lühr (GFZ Potsdam) for his encouragement and valuable comments on the manuscript. Many thanks to the DGG editorial team for their kind support in adapting the manuscript to the DGG publication format.

References:

• *ALLEY, R.B. & CLARK, P.U. (1999): The deglaciation of the northern hemisphere: a global perspective. – Annual Review of Earth and Planetary Science 27: 149–182.*

• CLARK, P.U., SHAKUN, J.D., BAKER, P.A., BARTLEIN, P.J., BREWER, S., BROOK, E., CARLSON, A.E., CHENG, H., KAUFMAN, D.S., LIU, Z., MARCHITTO, T.M., MIX, A.C., MORRILL, C., OTTOBLIESNER, B.L., PAHNKE, K., RUSSELL, J.M., WHITLOCK, C., ADKINS, J.F., BLOIS, J.L., CLARK, J., COLMAN, S.M., CURRY, W.B., FLOWER, B.P., HE, F., JOHNSON, T.C., LYNCH-STIEGLITZ, J., MARKGRAF, V., McMANUS, J., MITROVICA, J.X., MORENO, P.I. & WILLIAMS, J.W. (2012): Global climate evolution during the last deglaciation. – PNAS 109 (19): E1134-E1142.

• DOUGLAS, D.H. & CLADER, B.D. (2002): Climate sensitivity of the Earth to solar irradiance. – Geophysical Research Letters 29 (16): 33.1–33.4, doi: 10.1029/2002GL015345.

• LASKAR, J., JOUTEL, F. & BOUDIN, F. (1993): Orbital, precessional, and insolation quantities for the Earth from -20Myr to +10Myr. – Astronomy & Astrophysics 270: 522-533.

• PETIT, J.R. et al. (2001): Vostok Ice Core Data for 420,000 Years. – IGBP PAGES/World Data Center for Paleoclimatology Data Contribution Series #2001-076; Boulder, Colorado (NOAA/NGDC Paleoclimatology Program).

• SCHWARZ, O. (no date): Die Milankowitsch-Zyklen. – <www.physik.uni-siegen.de/didaktik/materialien_offen/milankowitsch.pdf>; Siegen (Universität Siegen) – Last access on August 7th, 2013.

• SHAKUN, J.D., CLARK, P.U., HE, F., LIFTON, N.A., LIU, Z. & OTTOBLISNER, B.L. (2015): Regional and global forcing of glacier retreat during the last deglaciation. – Nature Communications 6 (8059), doi: 10.1038/ncomms9059.

• *Wikimedia Commons (2013): File: Milankovitch Variations.png. <http://commons.wikimedia.org/wiki/File:Milankovitch_Variations.png>. Image created by ROHDE, R.A. – Last accesson September 15th, 2013.*

• *Wikipedia (2013): Milankovic'⊡-Zyklen. – Chapter „Änderung der Exzentrizität",<http://de.wikipedia.org/wiki/Milankovi%C4%87-Zyklen>. –*
Last access on September 16th, 2013.

There is no "Natural Atmospheric Greenhouse Effect"

Remarks: At the beginning, the author would like to remind you again of the "chicken trick" from the chapter "The beginnings of the climate religion", in which the power of the radiant heater is calculated down to the entire chicken including its back, which is the difference between a grilled (120°C) and a frozen (-18°C) chicken. In order to clarify the underlying misconception, you can simply test to refrain from turning the meet while grilling. Because according to the conventional S-B approach to the theoretical global temperature of our earth, a temperature medium from the top and bottom should appear on both sides of the meet...(sarcasm)

In the following, a two-layer case atmosphere / earth with 780W/m² temperature-effective solar radiation is considered, while in my blog articles a single-layer case (940W / m²) has been applied.

Scientific paper: WEBER, U.O. (2019): *„Weitere Überlegungen zur hemisphärischen Herleitung einer globalen Durchschnittstemperatur"* from the „Mitteilungen der Deutschen Geophysikalischen Gesellschaft" 1/2019: 18-25

Abstract

The Stefan-Boltzmann law (S-B law) describes for a black body the relationship between its specific temperature and the resulting radiation output in thermal equilibrium. The equal sign contained in the S-B law thus links clearly defined singular pairs of values and does not have the function of an arbitrary mathematical equation; it therefore does not apply to the ratio of arbitrarily determined average values.

Both inversions for calculating a theoretical global average temperature fail to meet this condition of the underlying S-B law: The conventional derivation from the earth's energy balance determines an average temperature from a globally averaged solar radiation of 235 W/m² and violates the strict condition for thermal equilibrium through the implicit inclusion of the night side of our earth. The hemispherical approach developed from this, with a temperature-effective net radiation from the sun of 390 W/m² heals the S-B equilibrium condition, but is also calculated using an average radiation.

A correct calculation of the theoretical global average temperature may only be derived from the individual local S-B equilibrium temperatures. As a result of this consideration, the temperature genesis on the day side of the earth can be explained by a refined hemispherical radiation approach with the Stefan-Boltzmann law, while the night cooling can be described with the ambient equation of the Stefan-Boltzmann law taking into account the heat content of the global circulations.

Note: The following considerations also implicitly include rounding errors that result from the use of different sources for the solar radiation quantities (WEBER 2016).

History

In the reports of the German Geophysical Society 3/2016 the article "A short note about the natural greenhouse effect" was published with a hemispherical Stefan-Boltzmann application for the calculation of a global average temperature (WEBER 2016). This hemispherical Stefan-Boltzmann approach was later presented in a simplified form on various internet platforms and discussed there in the comment functions. The author had explicitly pointed out in some of these later articles that his hemispherical derivation of the global average temperature with the Stefan-Boltzmann law can of course be scientifically refuted, quote:

„...If it were scientifically proven that the equalisation of the energy balance of our earth (surface of a sphere) with the strict thermal equilibrium requirement of the Stefan-Boltzmann law for the irradiated surface (hemisphere) is physically correct, then I am actually disproved...“

A closer look confirms the initial criticism of the conventional Stefan-Boltzmann approach for deriving a theoretical global average temperature. But the hemispherical S-B approach in its published form (WEBER 2016) also leads to a level of abstraction that is still too high and is cured with this work.

Critical examination of the hemispherical S-B approach for the global average temperature

On the internet platforms, the critical discussion was essentially limited to a fundamental rejection of the hemispherical S-B approach presented. Apart from the supposedly uncovered contradictions due to geometrical problems of understanding by individual commentators, a physically comprehensible falsification of this hemispherical S-B consideration had not occurred anywhere, especially not by the above-mentioned proof for the thermal equilibrium of a global energy balance. Rather, as evidence of the existence of an atmospheric greenhouse effect, there was argued with an atmospheric counter-radiation, which supposedly fills the difference between the calculated "S-B normal temperature" of our earth at -18°C and the actually measured average surface temperature of 14.8°C as commonly argued with (UBA 2013). Finally, a blog comment led to a lecture script by GERLICH (1995), where it says (quote):

„The radiation of a body depends on the actual temperature and not on any temperature averages! Average temperature values must always be determined from given temperature distributions and there are no solvable theoretical models for these average values. This clearly shows that all calculations with an "average radiation budget" or a "radiation balance" have nothing to do with average earth temperatures…"

Or to put it in other words: According to the legality clearly stipulated by the Stefan-Boltzmann law between the very specific temperature of a black body and its thereby clearly defined radiation power in a thermal equilibrium, there is no corresponding S-B average temperature value for any average amount of radiation energy. The following figures 1 and 2 of a S-B-compliant calculation for the day side of the earth illustrate the quoted statement by Gerlich (1995):

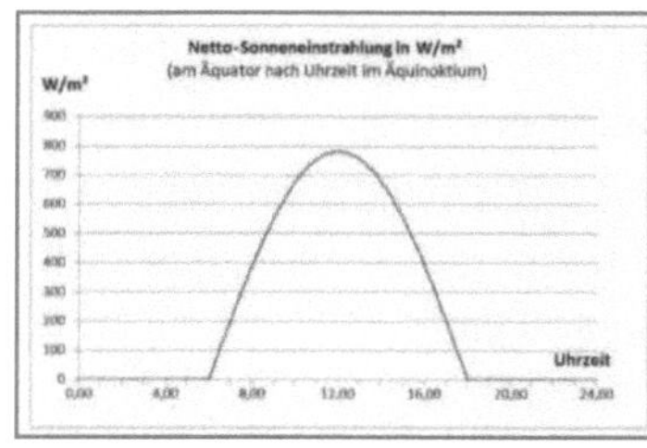

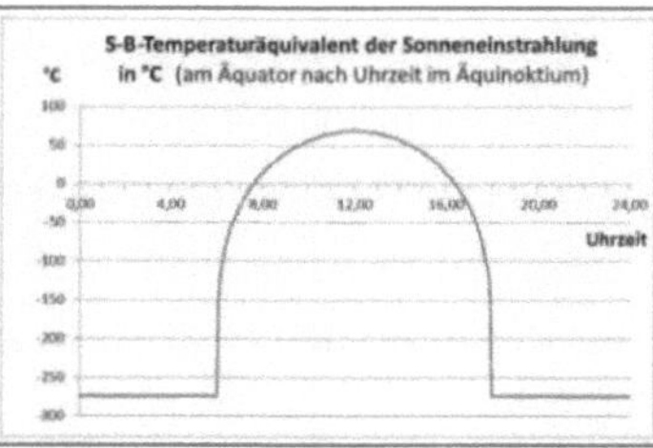

Abb. 1   **Abb. 2**

Note: The value of 780 W/m² is derived from WEBER (2016), where the temperature-effective solar radiation is viewed as a two-layer case for the atmosphere and the earth's surface. The equinox curve shown here for the time window between 6:00 a.m. and 6:00 p.m. also corresponds to the course of solar radiation between the two poles from -90 ° to + 90 ° latitude with the equator at 12:00 p.m. in the midday solar zenith.

In order to derive temperatures by means of an inversion of the Stefan-Boltzmann law in radiation equilibrium according to the formula $T = (S/\sigma)^{1/4}$, the following must basically be stated:

- The Stefan-Boltzmann law provides a physically proven unambiguous relationship between the specific temperature of a blackbody (primary) and its active radiation power (secondary) in a thermal equilibrium.

- The derivation of an induced temperature (secondary) from a passively obtained radiation power (primary), as used in both S-B approaches to determine the theoretical average temperature of the earth, requires the reversibility of the S-B law in thermal balance.

- Both S-B relationships, i.e. the S-B law itself and its inversion, deliver each an unambiguous calculative result in the thermal equilibrium for the specific radiation power assigned to an explicitly defined temperature or for the specific radiation power assigned to an explicitly defined temperature; both approaches therefore do not apply to average values.

With these statements, let us check the two discussed S-B inversions for calculating a global average temperature (Table 1):

Table 1: Comparison of S-B Inversions	Radiation Equilibrium	Averages
S-B Inversion over the global energy balance	No	Yes
Hemispherical S-B Inversion	Yes	Yes
A correct S-B Inversion were	Yes	No

Both S-B inversions for calculating the theoretical global average temperature initially require that an inversion of the Stefan-Boltzmann law is physically correct. But a theoretical average temperature of the earth results from the average of individual equilibrium temperatures and not as the result of an average radiation power:

- The **conventional S-B approach** is calculated using a globally averaged energy balance of 235 W/m², ignoring the mandatory implicit condition of the Stefan-Boltzmann law for thermal equilibrium due to the inclusion of the night side and is therefore too high abstraction. As a result, the global SB average temperature of -18°C, calculated on the basis of a global energy balance, can only provide a very rough "astronomical" minimum estimate and requires an explanation of the measured global average temperature of +15°C an additional atmospheric greenhouse effect for the difference of 33 degrees centigrade.

- The **hemispherical S-B approach** with an average hemispherical radiation output of 390 W/m² net is a significantly better approximation of the actual radiation conditions on the day side of the earth compared to the conventional approach. The hemispherical derived average S-B temperature agrees with the actual global measurement and does not require an atmospheric greenhouse effect. However, it also does not interpret the S-B law correctly because it derives a theoretical average temperature from a hemispherical averaged radiation power.

A T^4-relationship like the Stefan-Boltzmann law can therefore not depict any averages.

Example: 0 W/m² correspond to -273°C according to the S-B law and 470 W/m² correspond to +28°C. Their averaged temperature value of about -122.5°C for a radiation average of 235 W/m² does not correspond to the S-B temperature of -19°C, which is directly assigned to this radiation value of 235 W/m².

The equal sign in the Stefan-Boltzmann law thus creates a physically clear relationship between very specific radiation and temperature values in thermal equilibrium and should not be understood as any mathematical calculation instruction. The simultaneity on which the S-B law is based between concrete pairs of values for radiation and temperature would therefore be physically more clearly defined if this equal sign were replaced there by a double arrow, for example (equation 1):

(1) $\qquad P/A \Leftrightarrow \sigma * T^4$

$\qquad$ with the Stefan-Boltzmann-Constant $\sigma = 5{,}670 * 10^{-8}$ W m^{-2} K^{-4}, radiation P in W, area A in m² and temperature T in K.

The inversion of the Stefan-Boltzmann law would then look like this (Equation 2):

(2) $\qquad T \Leftrightarrow (S/\sigma)^{1/4} \qquad$ with $P/A = S$

The correct determination of an actual theoretical global average temperature must therefore be executed on the basis of individual local S-B equilibrium temperatures from the local latitude-dependent net solar radiation. Equation (6) from WEBER (2016) for a temperature-effective net radiation output of an average of 390 W/m² on the day side of the earth must therefore as equation (3) take into account the latitude dependence of the individual solar radiation forcing for the respective location:

(3) $S_{\varphi,Z} = 780\ [W/m^2] * \cos \varphi$

with the maximum latitude-dependent net radiation in the solar zenith $S_{\varphi,Z}$ and the latitude φ of the individual location corrected for the seasonal position of the sun. And for any average value from hemispherical maximum temperatures:

(4) $T = (\Sigma_{i=1\text{-}n}\ (780\ [W/m^2] * \cos \varphi_i / \sigma)^{1/4}) / n$

The average temperature of -90° to + 90° latitude above the individual S-B equilibrium temperatures for an equinoxial position of the sun at the zenith is then around 21°C.

Estimation for global temperature genesis

Figure 3 shows the different S-B temperature models during the day for an equatorial location in the equinox.

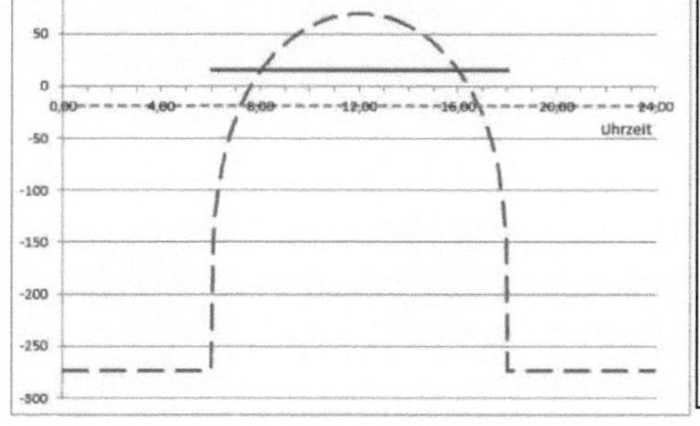

Figure 3: Calculative S-B equilibrium temperature in °C over the 24h-day in equinox

Blue / dashed: Local S-B equilibrium temperature at the equator

Red / solid: S-B temperature from hemispherical averaged solar radiation of 390 W/m²

Green / dotted: Conventional S-B average temperature from global energy balance of 235 W/m²

The temperature measured during the day for any location on the equator will be based on the course of the solar radiation and thus on the individual S-B equilibrium temperature and not on any globally determined average value; night cooling, on the other hand, depends on the available ambient heat. The calculated maximum S-B equilibrium temperature of almost 70°C exceeds the highest temperature ever measured of 57.7°C (Wikipedia 2017), so that an additional temperature effect does not appear to be necessary for the genesis of the individual local temperatures.

The night-time cooling of our earth is not correctly represented in both S-B approaches for the theoretical average temperature. In the hemispherical S-B approach, a nighttime radiation power of 0 W/m² is used, corresponding to an S-B temperature equivalent of -273.15°C, which, however, is not included in the calculation of the hemispherical daytime temperature. In the conventional approach, on the other hand, this nocturnal radiation power of 0 W/m² is implicitly included in the determination of the average temperature due to the global averaging.

On the latitude dependence of the local temperatures: Figure 4 shows the latitude-dependent annual course of the maximum S-B equilibrium temperature in the solar zenith in 20° increments of latitude.

It is immediately clear from the very different course of the curves that no statement whatsoever about the actual course of the current climate genesis on Earth can be derived from any type of a global average temperature. Rather, such global average temperature is a chimera that generalizes individual local changes in such a way that contrary trends can even become undetected in the end. Because of that, a reference to actual local changes, such as possible changes in the geographic climate zones, can no longer be assigned locally or even traced individually. Since the change in such an undifferentiated global temperature can no longer be traced back to specific local positions, a change in this global mean temperature can ultimately serve as a universal argument for any claim.

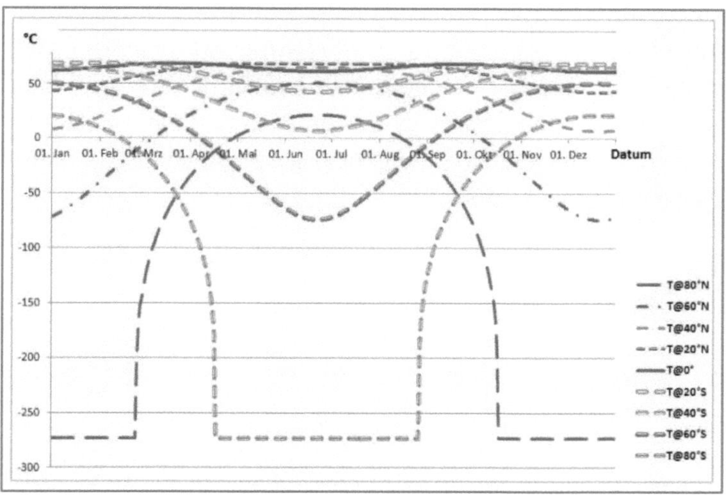

Figure 4: Latitude-dependent course of the maximum S-B temperature equivalent from insolation in °C through the year

Figure 4 clearly shows a high annual temperature constant around the equator and an increase in seasonal temperature effects with latitude. The S-B equilibrium temperatures in middle and higher latitudes change systematically with the seasonal position of the sun in both directions. Only in the respective winter half-year the maximum S-B temperature equivalent of the solar radiation above 40° latitude declines below 0°C, i.e. in the respective winter half-year on about one sixth of the earth's surface.

A comparison with the moon of our earth: Let us now consider the situation on the moon. The moon is a very simplified model of our earth without atmosphere and oceans.

VASAVADA et al. (2012) present measurements from the Diviner Lunar Radiometer Experiment (DLRE) for the surface temperature of the lunar surface in an equatorial position (Figure 5a). If you now compare the hemispherical Stefan-Boltzmann equilibrium temperature for the lunar equator (Figure 5b), you get a rather similar temperature curve:

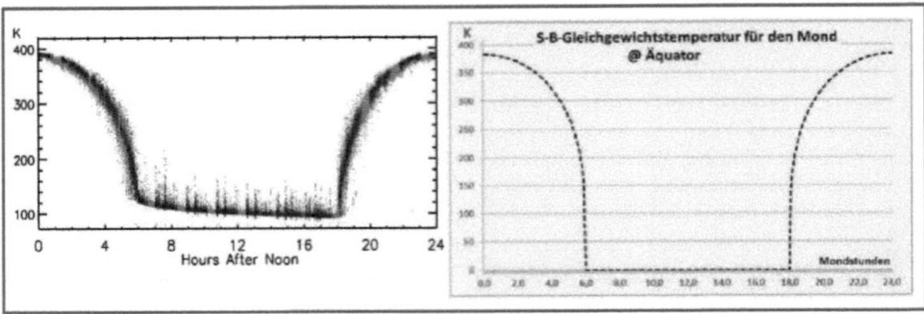

Figure 5: Surface temperature on the moon at the equator: a (left): Measured values from VASAVADA et al. (2012): Albedo greater than 0.13 (orange), less than 0.09 (green);
b (right): Equatorial hemispherical S-B maximum temperature

The hemispherical S-B calculation was performed without taking the lunar axis inclination into account with the following basic parameters:

Solar constant 1,367 W/m², Albedo of the moon 0.11, temperature effective solar radiation 1.217 W / m²,

effective solar radiation Seff for α = [0 ° - 360 °]:
IF COS α> 0 THEN Seff = COS α * 1,217 W / m²,
IF COS α <0 THEN Seff = 0 W / m².

If one projects the two figures 5a and 5b on top of each other, there is a very surprising agreement (figure 6):

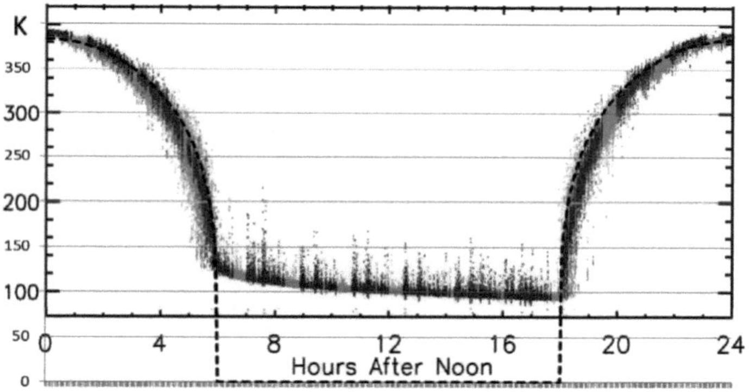

Figure 6 (Graphical combination of the two Figures 5a and 5b): Measured surface temperature on the moon (orange / green) and calculated S-B equilibrium temperature (dashed black)

The measured temperature and the hemispherical calculated S-B temperature for the equator of the moon match perfectly during the day, while the night temperatures differ by around 70 K. However, this difference is more of a problem between theory and practice, because in the hemispherical Stefan-Boltzmann calculation, radiation of 0 W/m² is strictly applied to the night side of the moon without storage of thermal energy. In fact, however, the temperature on the night side of the moon does not drop to 0 K due to the background radiation from space and the heat storage by the moon rock heated during the day.

This gives the moon an excellent match between the measured surface temperatures and the hemispherical determined S-B maximum temperatures. In addition to its surface, the moon has no additional heat storage. So the question arises for the hemispherical S-B model, how the night time drop of temperatures and the heat content of the global circulations could be taken into account.

The ambient equation for the Stefan-Boltzmann law: Let us consider the average global radiation of the earth. So far, we had considered the temperature-effective solar radiation of 780 W/m² from the two-layer case for atmosphere and earth surface postulated by WEBER (2016). When balancing the average amount of energy radiated by the earth at a constant mean temperature, we must now take into account the total amount of unreflected solar radiation. The solar constant minus the reflected portion accounts 940 W/m², the average for the entire earth's surface is 235 W/m². The Stefan-Boltzmann law for a black body in a heated environment (S-B ambient equation) is given by:

(5) $\qquad \Delta S = S - S_0 = \sigma * (T^4 - T_0^4)$ $\qquad$ with $S = P/A$ in W/m²

$\qquad$ (GERTHSEN & KNESER 1971)

Figure 7 shows that it is significant for the consideration of the global radiation at which temperature level we apply an average emission of $\Delta S = 235$ W/m². Because the temperature of a black body cannot fall below its ambient temperature T_0, which is determined here on Earth by the global circulations:

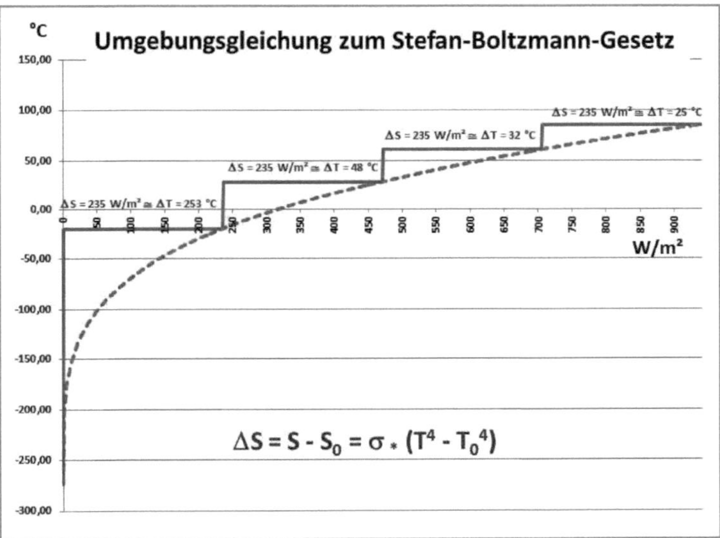

$$\Delta S = S - S_0 = \sigma * (T^4 - T_0^4)$$

Figure 7: The resulting temperature range for a global radiation of 235 W/m²: red or dashed: correlation between radiation and temperature according to the Stefan-Boltzmann law; blue or solid: S-B temperature equivalent for 235 W/m² depending on the respective ambient temperature

With the ambient equation of the S-B law, the only shortcoming of the hemispherical S-B model has now been cured, namely the lack of an approach for the local night temperature. Figure 7 clarifies that with the Stefan-Boltzmann law it is of crucial importance on which ambient temperature "T_0" a global radiation "ΔS" is based. The conventional S-B approach from the global energy balance assumes that the earth's ambient temperature is 0 K. From there, the known -19 °C is then calculated using the Stefan-Boltzmann law, as shown in Figure 7 by the first "step" from 0 to

235 W/m². The conventional S-B derivation for the "natural" global average temperature from the global energy balance thus ignores the heat content of the global circulations.

If we take the globally averaged measured average temperature (NST) of 14.8 °C for the ambient temperature T_0, which already implicitly includes the night temperature, then we have to divide the average global radiation of 235 W/m² from the global energy balance into the values -104 W/m² and +131 W/m² in order to correspond in their fluctuation range to the measured global average temperature of 14.8 ° C at 390 W/m². The arithmetical S-B equivalents result from this

$$390 \text{ W/m}^2 - 104 \text{ W/m}^2 = 286 \text{ W/m}^2 \cong -6{,}7 \text{ °C and}$$

$$390 \text{ W/m}^2 + 131 \text{ W/m}^2 = 521 \text{ W/m}^2 \cong +36{,}4 \text{ °C.}$$

The arithmetic mean of these two temperature values is +14.9 ° C and immediately leads to a contradiction: An average daily fluctuation range of around 43 °C is completely unrealistic for most areas on earth and may at best could be reached in fully arid desert areas of the lower geographic latitudes, i.e. precisely where the lowest heat inflow occurs from the global circulations. Conversely, a significant part of the intraday and winter temperature compensation is likely to take place via a participation of atmosphere and oceans according to the Stefan-Boltzmann ambient equation. The annual mean of the energy balance of the atmosphere and its components as a function of the geographic latitude shown in Figure 8 according to HÄCKEL (1990) clearly demonstrates this relationship.

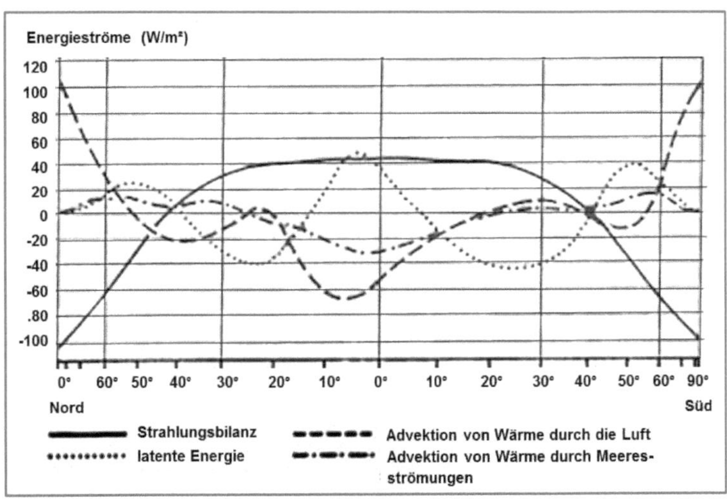

Figure 8: *Annual mean energy balance of the atmosphere and its components depending on the geographic latitude (according to HÄCKEL 1990)*

Figure 8 shows, based on a global energy balance of 235 W / m², the average annual heat flow due to global circulations from equatorial latitudes into middle and higher latitudes. In fact, these amounts are likely to be significantly higher in the respective winter half-year of the middle and higher latitudes than the annual average. Since the consideration with the ambient equation of the Stefan-Boltzmann law is also based on an average global radiation of 235 W/m², Figure 8 from Häckel fits into the present argument without contradiction.

This proves that the hemispherical determined radiation deficits in the respective winter half-year of the middle and higher latitudes are mitigated from the horizontal heat transport of the global circulations.

With the ambient equation of the S-B law, we can now include the influence of global circulations as the "ambience" of the earth in the consideration of the hemispherical temperature genesis. In particular, the oceans lose very little temperature overnight despite the lack of sunlight. The heat content of the global circulations alone ensures that the temperature level of the earth does not drop to near 0 K at night, as we can observe on the moon. The "T_0" in the S-B ambient equation should therefore be roughly represented by the globally averaged morning temperature of the oceans shortly before sunrise. The supply of heat energy in these circulations from the hemispherical solar radiation and nightly or winter inflows from these circulations smooth out the local temperature profile on earth compared to the hemispherical calculated maximum S-B radiation temperature.

Discussion

As has been shown, a hemispherical derivation of latitude-dependent local temperatures according to the Stefan-Boltzmann law needs no so-called atmospheric greenhouse effect. In contrast, the conventional derivation of a global average temperature from the average energy balance of the earth does not meet the implicit boundary conditions of the Stefan-Boltzmann law. Rather, it is incorrectly based on average values and, moreover, the required thermal equilibrium is missing between radiation and temperature. This conventional S-B derivation from the energy balance continues to suffer from an assumed night or ambient temperature of 0 K.

And finally, for a consideration of the global energy balance of the earth, it is not only the amount of an average radiation value of

235 W/m² that is decisive when it comes to a global averaging of the hemispherical supplied solar radiation over the total area of the earth, but that temperature level from which such radiation occurs.

At first glance, the temperature profiles on Earth and on the Moon are completely different. While the moon shows a very good agreement with the temperature curve of the calculated hemispherical S-B maximum temperature, the temperature on Earth is much more balanced and remains significantly below the hemispherical calculated minimum and maximum S-B extreme values. Earth and moon differ first of all by their day lengths of 24 hours and 29.5 earth days. The earth still differs from the moon in the thermal storage capacity of its global circulations (atmosphere and oceans). The short daily length of the earth and its heat storage therefore ensure little night cooling on earth, while the average night temperature on the moon quickly drops into the 3-digit minus range.

The heat stores of the global circulations alone prevent our earth from cooling down at night in the same order of magnitude as we can observe at the Moon. However, while the Moon without atmosphere and oceans roughly reaches the S-B equilibrium temperature on its day side, some of this energy is used on our earth to thermally charge the global circulations. And because of this heat storage, the minimum temperature of the earth drops again at most by a few decadegrees. Furthermore, these global circulations ensure in the respective hemispherical winter half-year that the local temperature in middle and higher latitudes cannot fall back to the pure S-B radiation temperature.

Natural fluctuations in these global circulation systems are then perceived regionally as climate changes, for example the El Niño phenomenon.

So if one wants to speak of a "natural temperature effect" on Earth, then this effect consists in damping the fluctuation range of locally measured temperatures compared to the respective hemispherical S-B equilibrium temperature. In the local radiation maximum energy is transferred to the global circulations, while at the local radiation minimum there is an inflow of heat from these circulations

Result

With this explanation of the Stefan-Boltzmann-law and the global temperature genesis it should be finally clarified that the warming of our Earth depends exclusively on the actual solar radiation on its day side. The S-B maximum temperatures determined from the radiation of the sun in terms of latitude lie between the tropics and, in the summer months, up to the middle latitudes of the respective hemisphere, significantly above the measured local temperatures. Here, between the tropics up to middle latitudes in the summer hemisphere, the global circulations are "charged" with thermal energy. At night and in the hemispherical winter half-year of the middle and higher latitudes, the heat content of these global circulations then ensures temperature compensation in relation to the low or absent effect of solar radiation. This nocturnal or winter cooling can be described with the ambient equation of the Stefan-Boltzmann law, taking into account the heat content of the global circulations.

The latitude-dependent derivation of individual local equilibrium temperatures from the hemispherical solar radiation of 780 W/m² net is thus a significantly improved model compared to the conventional Stefan-Boltzmann approach for an average temperature of our Earth from the global energy balance with 235 W/m². The hemispherical radiation model refutes the so-called atmospheric greenhouse effect and, in the form presented here as equation 3, fulfills all the conditions of the underlying Stefan-Boltzmann law.

Conclusion and Outlook

The result shown suggests that the correct determination of a theoretical global average temperature on the basis of the individual S-B equilibrium temperatures must be based on the actual latitude-dependent net solar radiation, and that analogously to equation (4) for all stations in the global temperature measurement network with application of the algorithms used for the measured average temperature. In such a calculation, the positive and negative heat flow between the respective location and the global circulations must be also taken into account.

Only with such a procedure a final comparison of the measured and theoretical temperature of our Earth could be carried out. However, such an end result should not only lead to a theoretical global average temperature as an undifferentiated measure for possible climate changes, but rather this result would have to be directly related to the geographic climate zones of our Earth. Only such a globally differentiated representation could lead to a meaningful visualization of the initial climatic situation on our Earth and its temporal changes.

Acknowledgements

I would like to thank the DGG editorial team for publishing my heretical article on the atmospheric greenhouse effect in their Mitteilungen 3/2016. For the courage to having made my critical statements public I have to thank the geologist and university professor Eckart Walger. In the colloquia and discussions at the CAU Kiel in the 1970s about the results of new geoscientific research, he was the one who regularly asked the very simple questions. These were the obvious questions that we students never dared to ask - and usually these questions could not be answered. Without this early and fundamental experience of critical science, I would certainly not have dared to publicly question the atmospheric greenhouse effect.

References

GERLICH, G. (1995): Die physikalischen Grundlagen des Treibhauseffektes und fiktiver Treibhauseffekte. – Manuskript zum Vortrag auf dem Herbstkongress der Europäischen Akademie für Umweltfragen: Die Treibhaus-Kontroverse, Leipzig, 9./10. November 1995; <www.ib-rauch.de/datenbank/vortrag-leipzig.html> (letzter Zugriff: 29.6.2017).

Gerthsen, C. & Kneser, H.O. (1971): Physik. – 11. Aufl.; Berlin (Springer); ISBN 3-54005562-2.

HÄCKEL, H. (1990): Meteorologie. – 8. Aufl. 2016; Stuttgart (Verlag Eugen Ulmer), ISBN 978-3-8252-4603-7.

UBA (2013): Wie funktioniert der Treibhauseffekt? – www.umweltbundesamt.de/service/uba-fragen/wie-funktioniert-der-treibhauseffekt (letzter Zugriff: 18.1.2018)

VASAVADA, A.R., Bandfield, J.L., Greenhagen, B.T., Hayne, P.O., Siegler, M.A., Williams, J.-P. & Paige, D.A. (2012): Lunar equatorial surface temperatures and regolith properties from the Diviner Lunar Radiometer Experiment. – J. Geophys. Res., 117: E00H18; doi: 10.1029/2011JE003987.

WEBER, U.O. (2016): A short note about the natural greenhouse effect. – Mitteilungen der Deutschen Geophysikalischen Gesellschaft 3/2016: 19-22.

Wikipedia (2017): El Azizia. – https://de.wikipedia.org/wiki/Al-%CA%BFAz%C4%ABz%C4%ABya (letzter Zugriff: 29.6. 2017).

After the Game is Before the Game*

* Citation: Sepp Herberger

(famous German national soccer coach from WM 1954)

The growing gap between individual knowledge and technological progress creates the inner space for a new superstition; and only for this reason the obsession of a man-made climate catastrophe is so widespread. Religion was always directed outward, into the unknown. The ancient pagan religions, the Greek, the Roman and the Nordic, were staged by their gods as an image of their world. The monotheistic religions and communism, on the other hand, had to create an inner antipode to their future promise of happiness and thus block out parts of reality, just like the CO_2 climate religion. Christian doctrine defines expulsion from paradise as the original sin of every single human who can only find salvation at the end of an earthly path of deprivation in a happy hereafter. Our technical civilization has now established this paradise on the basis of fossil fuels in the here and now, and the roots of our Christian occidental culture are consistently demanding repentance for our CO_2 inheritance, even if we have lost our religious believe.

The sentence quoted at the beginning, "This includes the fact that we do not call lies truth and truths do not lie!" (Merkel @ Harvard 2019) describes exactly the basic principle of science. But personal certainties have become more important in a hysterical environment as basic knowledge, and a few pieces of silver have turned into comfortable income at the expense of taxpayers. A reduced media spectrum announces to us today a targeted public opinion, which is driven by courtesy science and professional astroturfs as well as just-in-time actions by NGOs. We are celebrating a constitution that, in its politically correct interpretation, excludes competition with the socio-political opponent, while the recollection of the killing-order at the Berlin wall, the poor command economy and smelly lignite fires of a dilapidated, uniformly gray GDR are increasingly being superseded. Our gifted democracy had lasted just half a century and is now slowly creeping into a new ideological, unified society made up of eco-communist zealots for globalization and a wealth-indolent, silent majority. From the former sovereign of "mature citizens" it went down very quickly, via a "population" to those "who have been living here for a long time"; and one would like to shout out at this nonsense that a hunter also distinguishes his prey between seasonal and local game.

A CO_2 tax is the brilliant idea of a global elite to tax breathing and thus life itself - and thereby overrule the free market economy in a planned manner. According to the wishes of pubescent children, this "climate protection" should be included in the Basic Law in the future. And then it will probably only be a matter of time before the so-called "climate deniers" will eventually be persecuted as "counter-revolutionaries" or "climate pests" and at least their books will burn.

The scientific knowledge presented in this book did not even make it to the (German) public eye, but got stuck in the filter bubble of the so-called climate realists. And that is why the climate hysterics were not forced to substantiate their false doctrine. With rather cautious approval, the so-called climate realists mostly delivered non-factual comments, up to ad-hominem contributions from academic researchers and professors, while there was often proposed averaging in a physical exponent-4 law and an extremely self-confident STEM weakness became visible in spatial imagination. And many of those whose own insights point in a similar direction to that of the author could not take the mental step from the Stefan-Boltzmann law to its ambient equation, were not able to overcome choleric outbursts despite their level of knowledge, could not withhold from abstruse allegations of plagiarism or have been implicitly silent.

The critics of the climate alarm are certainly all very tough guys who tend to think independently and do not allow a "X" to be faked for an "U". On the other hand, most of them are not team players or strategists who act collectively against the climate mania, but stubborn one-hand-sailors who only react with their own resources.

These so-called climate realists now represent the rational counterpart to those climate-religious scientists who want to steer our state ship with "full force ahead" against the iceberg of a global ecological transformation on which the world as we know it today has to shatter. And all climate realists agree that a "full force back" maneuver should take place now; however, they cannot agree between each other on some degrees to the port or starboard side and therefore quarrel like the tinkerers.

But even so, these voluntary critics of the climate frenzy, who fervently disassemble each other in their filter bubble, are much more seekers of truth than the entire established climate science.

Finally, please follow the author for a brief review of our cultural evolution, which ultimately led to the blessings of our industrial age. The amount of energy available per capita had increased drastically in several revolutionary steps, and all of these transitions had developed in free market competition to the previously existing system:

- Stone Age (= small local communities of hunters and food gatherers): The amount of energy available per capita was about 3 to 6 times the basic human requirement.

- Age of agriculture and livestock breeding (= regional civilizations): The amount of energy available per capita was around 18 to 24 times the basic human requirement.

- Industrial age (= globalized world): The amount of energy available per capita today is around 70 to 80 times the basic human requirement.

Only the industrial use of fossil fuels since the beginning of industrialization has raised our standard of living, the availability and quality of food, the daily working hours, the free weekend, the annual vacation entitlement, the healthcare system, individual life expectancy, transportation, the communication and the general technological standard to what we now consider to be "quite normal". We all live today in the way that only the Roman emperors could afford two thousand years ago, thanks to fossil fuels, the latest technologies and energy available at all times. Conversely, this means that our current standard of living - and also our current welfare - is based on the use of such fossil fuels; because only an added value can be distributed without consuming the substance.

The planned ideological turns, i.e. energy turnaround, traffic turnaround and agricultural turnaround, cannot achieve any real added value in an economic sense and consequently lead directly to a subsidized planned economy; their added value is purely emotional. The following limit value analysis shows which problems our society is facing today.

The limit for a subsidized planned economy: A subsidized planned economy in favor of global climate protection is not subject to any economic competition with alternative solutions. Rather, a given climate-related measure is being promoted through subsidies because the average consumer has no alternative choice at all. But because planned economy eliminates bankruptcy as an economic regulation, all wrong decisions and bad planning must also be compensated by subsidies in future. Due to the system, it is no longer possible to reduce such subsidies because a planned economy carries no alternative solutions inside. Rather, a subsidized planned economy à la Eastern Bloc must inevitably lead to a corresponding reduction in quality, innovation and supply - and, ultimately, can no longer meet the demand of a deficient society.

The limit for emotional added value: The climate protectionists' two-earth argumentation demands that all believers preserve our natural environment and the global climate for future generations. Active participation in the implementation of this high ecological goal provides everyone involved, in addition to economic security, in particular emotional feelings of happiness and confirmation of their religiously excessive self-image. Ultimately, this exaggeration no longer allows one to deal with people who think differently about alternatives and must therefore lead directly to ecological totalitarianism via a "creative state".

Synthesis of these two limits: A subsidized planned economy must converge in a socialist deficit economy because there is no economic regulation of free competition. This turns the two-earth argument into its opposite, because two earths would also be necessary for an ecologically correct nutrition of the entire world population. The increased consumption of land for the faithful production of food and energy must therefore end up in an economic supply disaster. And because the human mentality to survive is much more pronounced than any idea of environmental protection, ecototalitarism will collapse at some point. And the survivors of this catastrophe will eventually have to relentlessly plunder the planet's ecological resources, just to survive.

Conclusion: The decarbonization of the world is not a cultural advance for mankind, but leads directly back to an ecological Middle Ages society with about one quarter of the currently available energy per capita. And in this brave new world, our descendants will eat their bread in the sweat of their faces again and will dream of a lost industrial paradise...

Paperback B/W - 48 pp German

ISBN-13: 9783752870343

E-Book 3,49 € - Book 4,99 €

The hemispheric Stefan-Boltzmann approach does not deal with weather or climate, but a long-term mean of the elementary forces that underlie both. Because in the balance between hemispherical irradiation of the sun and global radiation from the Earth, the terrestrial heat stores of atmosphere and oceans determine the individual local temperature profile on Earth. A "natural atmospheric greenhouse effect" or an "atmospheric counter-radiation" are neither necessary nor demonstrable to explain the temperature genesis on our earth.

Paperback B/W - 128 pp German

ISBN-13: 9783752804355

E-Book 3,99 € - Book 5,99 €

The perspective to understand transformations in our country (Germany) has changed diametrically. The current change of society is not commented from a conservative middle, as we had experienced during our so-called '68 student revolution, but rather this conservative social center itself, allegedly as "old white globalization losers", is the subject of a devastating public criticism.